21世纪高等院校计算机网络工程专业规划教材

路由交换技术实践教程

孙良旭 尹航 主编
杨丹 董立文 王刚 编著

清华大学出版社
北京

内 容 简 介

本书共 11 章，以 Cisco 的中低端路由器和交换机为核心，全面介绍路由交换技术，内容包括 IOS 配置基础、接口与管理配置、IP 特性配置、广域网配置、网络安全配置、动态路由协议配置、交换机基本配置、虚拟局域网、生成树、虚拟网间路由和综合案例。

本书内容丰富，实例众多，图文并茂，结构合理，适合作为高等院校的教材，也可以供网络工程技术人员参考。

本书封面贴有清华大学出版社防伪标签，无标签者不得销售。
版权所有，侵权必究。举报：010-62782989，beiqinquan@tup.tsinghua.edu.cn。

图书在版编目（CIP）数据

路由交换技术实践教程/孙良旭等主编. —北京：清华大学出版社，2014（2025.2 重印）
（21 世纪高等院校计算机网络工程专业规划教材）
ISBN 978-7-302-34787-3

Ⅰ. ①路… Ⅱ. ①孙… Ⅲ. ①计算机网络－路由选择－高等学校－教材 ②计算机网络－信息交换机－高等学校－教材 Ⅳ. ①TN915.05

中国版本图书馆 CIP 数据核字(2013)第 301267 号

责任编辑：付弘宇　薛　阳
封面设计：何凤霞
责任校对：时翠兰
责任印制：刘　菲

出版发行：清华大学出版社
　　　网　　址：https://www.tup.com.cn，https://www.wqxuetang.com
　　　地　　址：北京清华大学学研大厦 A 座　　邮　　编：100084
　　　社 总 机：010-83470000　　　　　　　　邮　　购：010-62786544
　　　投稿与读者服务：010-62776969，c-service@tup.tsinghua.edu.cn
　　　质量反馈：010-62772015，zhiliang@tup.tsinghua.edu.cn
　　　课件下载：https://www.tup.com.cn，010-83470236
印 装 者：涿州市般润文化传播有限公司
经　　销：全国新华书店
开　　本：185mm×260mm　　印　张：12.5　　字　数：305 千字
版　　次：2014 年 4 月第 1 版　　　　　　　印　次：2025 年 2 月第 9 次印刷
定　　价：39.80 元

产品编号：056145-02

前　言

随着网络技术的发展,网络工程技术人员越来越受到社会的欢迎。我校的网络工程专业是我国高校在此专业的首批试点单位,本专业的"路由交换技术"课程是一门实践性很强的课程,其实践教学无论在授课内容、授课学时,还是在授课方式上都没有先例可循。

本教材在内容上设计多个知识点案例和一个综合案例,注重工程实践应用背景,意在启发和引导学生能将所学的路由与交换技术原理及技术应用付诸于实践,帮助学生真正掌握安装、配置、调试和运营局域网及广域网所需的实践技能,做到学以致用。

思科(Cisco)是目前全球领先的网络设备和解决方案供应商,其网络设备和解决方案在我国已广泛应用,并得到认可。交换机和路由器是组建网络基础设施的基石。本教材重点以 Cisco 的中低端的 Catalyst 交换机和路由器作为讲授的核心设备。

本教材具有如下特点:

(1) 知识点覆盖面广。知识点涵盖从数据链路层到应用层的重要网络配置需求。

(2) 重点突出。重点讲解基础的、实用的和主流的路由和交换技术的实践。

(3) 知识点案例与综合案例相结合。教材知识点案例阐述准确、清晰和简洁,综合案例设计典型、可操作和贴近实际。

(4) 结构完整统一,设计合理科学。教材每章包括基础知识和知识点案例实验两大部分,基础知识部分简单扼要地讲解相关理论知识,知识点案例实验部分重点从多个不同的知识点讲解相关的技术应用。最后教材设计综合案例章节选取校园网典型案例,综合应用前面各章内容完成实际案例配置需求。

(5) 例题讲解配置完整,解释清晰。全书案例都提供完整的拓扑图、命令配置、测试输出和解释说明,使读者知其然,更知其所以然。

(6) 本教材作为《路由交换技术》教材配套的实践教材,适合作为不同专业、不同层次的教材或者工程技术参考。

由于作者水平有限,错误或不当之处在所难免,殷切期望读者批评指正。

作　者

2014 年 1 月于辽宁科技大学

目　录

第 1 章　IOS 配置基础 1

　1.1　基础知识 1

　　　1.1.1　IOS 概述 1

　　　1.1.2　基本硬件构件 1

　　　1.1.3　基本软件构件 2

　　　1.1.4　IOS 配置方法 2

　　　1.1.5　命令行的使用 3

　1.2　实验 1　控制台方式访问路由器 4

　　　1.2.1　实验拓扑 4

　　　1.2.2　实验内容 4

　　　1.2.3　实验步骤 4

第 2 章　接口与管理配置 9

　2.1　基础知识 9

　　　2.1.1　接口配置 9

　　　2.1.2　系统管理 9

　　　2.1.3　文件管理 9

　　　2.1.4　故障处理 10

　　　2.1.5　CDP 介绍 10

　2.2　实验 1　路由器基本配置 10

　　　2.2.1　实验拓扑 10

　　　2.2.2　实验内容 11

　　　2.2.3　实验步骤 11

　2.3　实验 2　路由器高级配置命令 13

　　　2.3.1　实验拓扑 13

　　　2.3.2　实验内容 13

　　　2.3.3　实验步骤 13

　2.4　实验 3　路由器配置文件备份 15

　　　2.4.1　实验拓扑 15

　　　2.4.2　实验内容 15

	2.4.3	实验步骤	15
2.5	实验4	路由器的密码设置、保存与破解方法	17
	2.5.1	实验拓扑	17
	2.5.2	实验内容	17
	2.5.3	实验步骤	17
2.6	实验5	CDP	18
	2.6.1	实验拓扑	18
	2.6.2	实验内容	19
	2.6.3	实验步骤	19

第3章　IP特性配置 22

3.1	基础知识		22
	3.1.1	地址解析方法	22
	3.1.2	广播包处理	22
	3.1.3	自治系统	22
	3.1.4	路由技术	23
	3.1.5	路由表	23
	3.1.6	管辖距离	23
	3.1.7	度量值	23
	3.1.8	路由更新	23
	3.1.9	路由查找	24
	3.1.10	静态路由和动态路由	24
	3.1.11	默认路由	24
	3.1.12	汇总路由	25
	3.1.13	DHCP概述	25
3.2	实验1	路由器的背对背连接	25
	3.2.1	实验拓扑	25
	3.2.2	实验内容	26
	3.2.3	实验步骤	26
3.3	实验2	静态路由协议	27
	3.3.1	实验拓扑	27
	3.3.2	实验内容	27
	3.3.3	实验步骤	27
3.4	实验3	默认路由	29
	3.4.1	实验拓扑	29
	3.4.2	实验内容	29
	3.4.3	实验步骤	29
3.5	实验4	DHCP基本配置	30
	3.5.1	实验拓扑	30

    3.5.2 实验内容 ·· 30

    3.5.3 实验步骤 ·· 30

  3.6 实验5 DHCP中继配置 ·· 32

    3.6.1 实验拓扑 ·· 32

    3.6.2 实验内容 ·· 32

    3.6.3 实验步骤 ·· 32

第4章 广域网配置 ·· 35

  4.1 基础知识 ·· 35

    4.1.1 DDN概述 ·· 35

    4.1.2 HDLC概述 ·· 35

    4.1.3 PPP概述 ·· 35

    4.1.4 帧中继概述 ·· 36

  4.2 实验1 HDLC和PPP封装 ·· 37

    4.2.1 实验拓扑 ·· 37

    4.2.2 实验内容 ·· 37

    4.2.3 实验步骤 ·· 37

  4.3 实验2 PAP认证实验 ·· 39

    4.3.1 实验拓扑 ·· 39

    4.3.2 实验内容 ·· 39

    4.3.3 实验步骤 ·· 40

  4.4 实验3 CHAP认证实验 ·· 41

    4.4.1 实验拓扑 ·· 41

    4.4.2 实验内容 ·· 41

    4.4.3 实验步骤 ·· 41

  4.5 实验4 路由器配置为帧中继交换机 ·· 42

    4.5.1 实验拓扑 ·· 42

    4.5.2 实验内容 ·· 42

    4.5.3 实验步骤 ·· 42

  4.6 实验5 帧中继点到点形式及帧中继映射 ·· 44

    4.6.1 实验拓扑 ·· 44

    4.6.2 实验内容 ·· 44

    4.6.3 实验步骤 ·· 44

  4.7 实验6 帧中继点到多点子接口 ·· 47

    4.7.1 实验拓扑 ·· 47

    4.7.2 实验内容 ·· 48

    4.7.3 实验步骤 ·· 48

第 5 章 网络安全配置 51

5.1 基础知识 51
5.1.1 ACL 概述 51
5.1.2 ACL 在网络中的应用位置 51
5.1.3 NAT 概述 52

5.2 实验 1 标准 ACL 实验 53
5.2.1 实验拓扑 53
5.2.2 实验内容 53
5.2.3 实验步骤 53

5.3 实验 2 扩展访问控制列表 55
5.3.1 实验拓扑 55
5.3.2 实验内容 56
5.3.3 实验步骤 56

5.4 实验 3 命名访问控制列表 58
5.4.1 实验拓扑 58
5.4.2 实验内容 59
5.4.3 实验步骤 59

5.5 实验 4 访问控制列表综合应用 60
5.5.1 实验拓扑 60
5.5.2 实验内容 61
5.5.3 实验步骤 61

5.6 实验 5 静态 NAT 服务 64
5.6.1 实验拓扑 64
5.6.2 实验内容 64
5.6.3 实验步骤 65

5.7 实验 6 动态 NAT 与 PAT 服务 66
5.7.1 实验拓扑 66
5.7.2 实验内容 67
5.7.3 实验步骤 67

第 6 章 动态路由协议配置 71

6.1 基础知识 71
6.1.1 RIP 概述 71
6.1.2 RIP 认证 71
6.1.3 自动汇总 71
6.1.4 IGRP 概述 72
6.1.5 IGRP 的度量权重 72
6.1.6 OSPF 概述 72

		6.1.7 OSPF 认证 ………………………………………………………… 72
		6.1.8 EIGRP 概述 ………………………………………………………… 73
		6.1.9 EIGRP 认证 ………………………………………………………… 73
	6.2	实验 1 RIP v1 路由协议配置 …………………………………………………… 73
		6.2.1 实验拓扑 ………………………………………………………… 73
		6.2.2 实验内容 ………………………………………………………… 74
		6.2.3 实验步骤 ………………………………………………………… 74
	6.3	实验 2 RIP v2 路由协议配置 …………………………………………………… 76
		6.3.1 实验拓扑 ………………………………………………………… 76
		6.3.2 实验内容 ………………………………………………………… 76
		6.3.3 实验步骤 ………………………………………………………… 76
	6.4	实验 3 RIP v2 认证和触发更新 ……………………………………………… 79
		6.4.1 实验拓扑 ………………………………………………………… 79
		6.4.2 实验内容 ………………………………………………………… 80
		6.4.3 实验步骤 ………………………………………………………… 80
	6.5	实验 4 EIGRP 基本配置 ……………………………………………………… 82
		6.5.1 实验拓扑 ………………………………………………………… 82
		6.5.2 实验内容 ………………………………………………………… 82
		6.5.3 实验步骤 ………………………………………………………… 82
	6.6	实验 5 EIGRP 路由汇总 ……………………………………………………… 86
		6.6.1 实验拓扑 ………………………………………………………… 86
		6.6.2 实验内容 ………………………………………………………… 86
		6.6.3 实验步骤 ………………………………………………………… 86
	6.7	实验 6 EIGRP 认证 …………………………………………………………… 87
		6.7.1 实验拓扑 ………………………………………………………… 87
		6.7.2 实验内容 ………………………………………………………… 88
		6.7.3 实验步骤 ………………………………………………………… 88
	6.8	实验 7 EIGRP 基本配置 ……………………………………………………… 89
		6.8.1 实验拓扑 ………………………………………………………… 89
		6.8.2 实验内容 ………………………………………………………… 90
		6.8.3 实验步骤 ………………………………………………………… 90
	6.9	实验 8 点到点链路 OSPF 配置 ……………………………………………… 94
		6.9.1 实验拓扑 ………………………………………………………… 94
		6.9.2 实验内容 ………………………………………………………… 94
		6.9.3 实验步骤 ………………………………………………………… 95
	6.10	实验 9 广播多路访问链路上的 OSPF ……………………………………… 99
		6.10.1 实验拓扑 ………………………………………………………… 99
		6.10.2 实验内容 ………………………………………………………… 99
		6.10.3 实验步骤 ………………………………………………………… 100

6.11 实验 10 基于区域的 OSPF 简单口令认证 …………………………………… 102
    6.11.1 实验拓扑 …………………………………………………………… 102
    6.11.2 实验内容 …………………………………………………………… 102
    6.11.3 实验步骤 …………………………………………………………… 102
6.12 实验 11 基于区域的 OSPF MD5 认证 ………………………………………… 104
    6.12.1 实验拓扑 …………………………………………………………… 104
    6.12.2 实验内容 …………………………………………………………… 105
    6.12.3 实验步骤 …………………………………………………………… 105

第 7 章 交换机基本配置 …………………………………………………………… 107

7.1 基础知识 ………………………………………………………………………… 107
    7.1.1 交换机工作原理 …………………………………………………… 107
    7.1.2 交换机功能 ………………………………………………………… 107
    7.1.3 交换机工作特性 …………………………………………………… 107
    7.1.4 交换机分类 ………………………………………………………… 107
7.2 实验 1 控制台方式访问交换机 ……………………………………………… 108
    7.2.1 实验拓扑 …………………………………………………………… 108
    7.2.2 实验内容 …………………………………………………………… 108
    7.2.3 实验步骤 …………………………………………………………… 108
7.3 实验 2 交换机基本配置 ……………………………………………………… 111
    7.3.1 实验拓扑 …………………………………………………………… 111
    7.3.2 实验内容 …………………………………………………………… 111
    7.3.3 实验步骤 …………………………………………………………… 111
7.4 实验 3 Telnet 方式访问交换机 ……………………………………………… 115
    7.4.1 实验拓扑 …………………………………………………………… 115
    7.4.2 实验内容 …………………………………………………………… 116
    7.4.3 实验步骤 …………………………………………………………… 116
7.5 实验 4 MAC 地址管理 ………………………………………………………… 118
    7.5.1 实验拓扑 …………………………………………………………… 118
    7.5.2 实验内容 …………………………………………………………… 118
    7.5.3 实验步骤 …………………………………………………………… 119

第 8 章 虚拟局域网 ……………………………………………………………… 125

8.1 基础知识 ………………………………………………………………………… 125
    8.1.1 VLAN 定义 ………………………………………………………… 125
    8.1.2 VLAN 的优点 ……………………………………………………… 126
    8.1.3 VLAN 划分 ………………………………………………………… 126
    8.1.4 VLAN 标准 ………………………………………………………… 127
8.2 实验 1 静态 VLAN …………………………………………………………… 127

    8.2.1 实验拓扑 ····· 127
    8.2.2 实验内容 ····· 127
    8.2.3 实验步骤 ····· 127
  8.3 实验2 干道链接 ····· 130
    8.3.1 实验拓扑 ····· 130
    8.3.2 实验内容 ····· 131
    8.3.3 实验步骤 ····· 131
  8.4 实验3 干道协议 ····· 138
    8.4.1 实验拓扑 ····· 138
    8.4.2 实验内容 ····· 138
    8.4.3 实验步骤 ····· 138

第9章 生成树 ····· 145

  9.1 基础知识 ····· 145
    9.1.1 STP概述 ····· 145
    9.1.2 STP算法 ····· 145
    9.1.3 网桥协议数据单元 ····· 146
    9.1.4 端口的状态 ····· 146
    9.1.5 选举根网桥 ····· 147
    9.1.6 选举根端口 ····· 147
    9.1.7 选举指定端口 ····· 147
  9.2 实验1 生成树 ····· 148
    9.2.1 实验拓扑 ····· 148
    9.2.2 实验内容 ····· 148
    9.2.3 实验步骤 ····· 148

第10章 虚拟网间路由 ····· 161

  10.1 基础知识 ····· 161
    10.1.1 一般路由器实现VLAN间路由 ····· 161
    10.1.2 单臂路由器实现VLAN间路由 ····· 161
    10.1.3 三层交换机实现VLAN间路由 ····· 163
  10.2 实验1 一般路由器实现VLAN间路由 ····· 163
    10.2.1 实验拓扑 ····· 163
    10.2.2 实验内容 ····· 164
    10.2.3 实验步骤 ····· 164
  10.3 实验2 单臂路由实现VLAN间路由 ····· 166
    10.3.1 实验拓扑 ····· 166
    10.3.2 实验内容 ····· 166
    10.3.3 实验步骤 ····· 166

10.4 实验3 三层交换实现VLAN间路由 ………………………………………… 168
 10.4.1 实验拓扑 ………………………………………………………… 168
 10.4.2 实验内容 ………………………………………………………… 168
 10.4.3 实验步骤 ………………………………………………………… 168

第11章 综合案例 ……………………………………………………………… 170

11.1 案例需求 …………………………………………………………………… 170
 11.1.1 案例背景和目的 ………………………………………………… 170
 11.1.2 案例配置需求 …………………………………………………… 170

11.2 拓扑图设计 ………………………………………………………………… 171
 11.2.1 拓扑图 …………………………………………………………… 171
 11.2.2 IP地址规划 ……………………………………………………… 171

11.3 需求配置 …………………………………………………………………… 172
 11.3.1 帧中继网络 ……………………………………………………… 172
 11.3.2 访问控制 ………………………………………………………… 174
 11.3.3 网络地址转换 …………………………………………………… 174
 11.3.4 动态地址分配 …………………………………………………… 175
 11.3.5 动静态路由 ……………………………………………………… 175
 11.3.6 VLAN间路由 …………………………………………………… 176
 11.3.7 STP负载分担 …………………………………………………… 178

11.4 设备配置 …………………………………………………………………… 179

参考文献 ………………………………………………………………………… 187

第1章 IOS 配置基础

1.1 基础知识

1.1.1 IOS 概述

互联网操作系统（Internetwork Operating System，IOS）是思科私有的核心软件数据包，主要在思科路由器和交换机上实现。Cisco IOS 软件通过一组增值技术和特性而具有因特网智能作用。Cisco IOS 包含一组精湛的网络技术，主要由互联网设备诸如路由器、交换机、PC 和工作站等提供。

Cisco 的网络设备需要依靠 IOS 这个操作系统进行工作，它指挥和协调 Cisco 设备的硬件进行网络服务和应用的传递。通过使用 IOS 命令，可以为 Cisco 网络设备进行各种各样的配置，使之适应于各种网络功能。

1.1.2 基本硬件构件

Cisco 路由器系列包含各种类型的路由产品，尽管这些产品的处理能力和所支持的接口数目具有相当大的差异，但它们都由相似的核心硬件构件所组成。尽管中央处理器（CPU）、只读存储器（ROM）和随机存取存储器（RAM）的数目及所使用的接口、介质转换器的数量和方式会因产品类别的差异而不同，但每台路由器均含有如图 1.1 所示的硬件构件，其中 2600 系列路由器内部构成如图 1.2 所示。

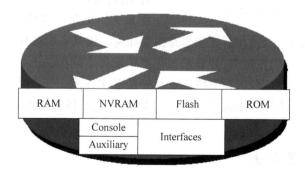

图 1.1 路由器基本硬件构件

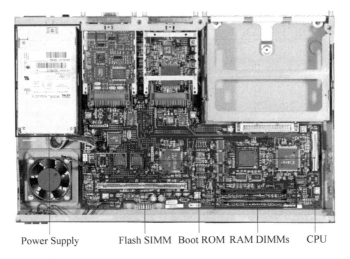

图 1.2　2600 系列路由器硬件构件

1.1.3　基本软件构件

Cisco 路由器有两种主要的软件构件：IOS 的映像文件和配置文件。这两种主要的路由器软件构件与路由器内存的对应关系如图 1.3 所示。

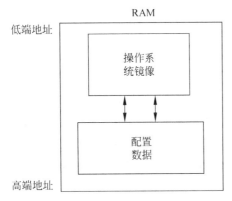

图 1.3　路由器基本软件构件

1. IOS 的映像文件

IOS 的映像是保存在存储器的 IOS 代码。映像文件以二进制形式保存。自启加载器根据配置寄存器所设定的内容定位操作系统镜像文件的位置，一旦找到镜像文件，便将其加载到内存的低端地址。操作系统的镜像文件包含一系列规则，这些规则规定如何通过路由器传送数据、管理缓存空间、支持不同的网络功能、更新路由表和执行用户命令。

2. 配置文件

配置文件由路由器管理员创建，其中存放的配置内容由操作系统解释，操作系统指示路由器如何完成其中的各种功能。例如，配置文件可以定义一个或多个访问控制表，并要求操作系统设置不同的访问控制表来访问不同的接口，以提供流入该路由器的包的控制级别。尽管配置文件定义了如何完成影响路由器运行的各种功能，实际上是由操作系统来完成这些工作的，这是因为操作系统解释并响应配置文件中所陈述的要求。

1.1.4　IOS 配置方法

路由器没有键盘和鼠标，要初始化路由器，需要把计算机的串口和路由器的 Console 口进行连接。访问 Cisco 路由器的方法还有 Telnet、Web browser、网管软件（如 Cisco Works）等。通常可以通过 5 种方法实现配置，如图 1.4 所示。

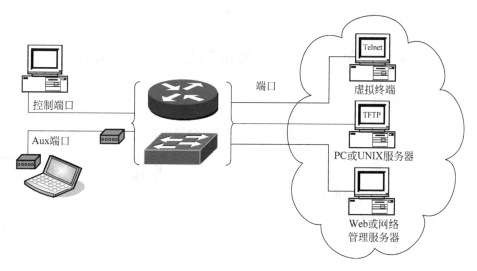

图 1.4　IOS 配置方式

1. 通过 Console 接口直接配置

将主机的串口通过 Console 线连接到设备的 Console 接口，并且使用主机的超级终端对设备进行配置。

2. 通过 Aux 接口进行远程配置

通过 Modem 连接设备的 Aux 接口，从远程通过拨号的方式配置设备。

3. 通过 Telnet 进行远程配置

通过 Telnet 远程登录到设备上进行配置。

4. 通过 TFTP 服务器进行远程配置

在网络中建立 TFTP 服务器，把设备的 IOS 和启动配置备份到该服务器，在必要时可以把该服务器上的备份恢复到设备中。

5. 通过 Web 或者 SNMP 网管工作站远程配置

通过启用 Web 配置方式，以使管理人员能够从远程通过浏览器来配置设备。由于 HTTP 的不安全性，不建议采用这样的方法配置路由器。通过网管工作站进行配置，这就需要在网络中有至少一台运行 Cisco Works 及 Cisco View 等的网管工作站。

IOS 默认方式是从控制台终端输入命令，这种方式是用户的主要配置方式。如果通过其他方式访问，使用者应首先配置路由器或交换机以便能使用该访问方式。

1.1.5　命令行的使用

Cisco 的 IOS 是命令行界面(Command Line Interface, CLI)。CLI 有两种基本工作模式：

（1）用户模式(User mode)。通常用来查看路由器的状态。在此状态下，无法对路由器进行配置，可以查看的路由器信息也是有限的。

（2）特权模式(Privilege mode)。可以更改路由器的配置，当然也可以查看路由器的所有信息。

除此之外，还具有全局配置模式、接口配置模式、子接口配置模式、线路配置模式、ROM

监控模式、路由器配置模式和 VLAN 配置模式等。

1.2 实验 1 控制台方式访问路由器

1.2.1 实验拓扑

一台 PC 通过 RS-232 串口使用控制台线连接一台路由器的 Console 端口,见图 1.5。

1.2.2 实验内容

(1) 通过控制台线缆实现 PC 连接路由器。
(2) 通过控制台程序实现 PC 访问路由器。

图 1.5 控制台方式连接路由器

1.2.3 实验步骤

(1) 选择【开始】→【所有程序】→【附件】→【通讯】→【超级终端】菜单项,启动超级终端程序,见图 1.6。

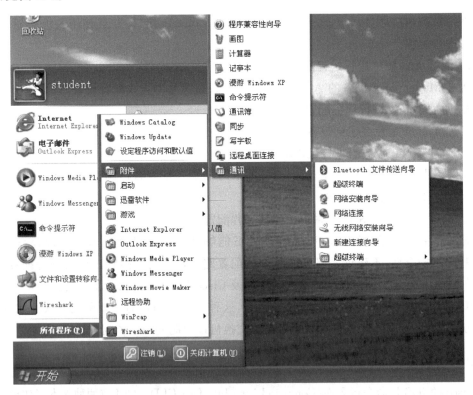

图 1.6 启动超级终端程序

(2) 输入新建连接名称为 cisco,选择任一图标,单击【确定】按钮,见图 1.7。
(3) 根据在连接时使用列表选择 COM1,单击【确定】按钮,见图 1.8。
(4) 单击【还原为默认值】按钮,还原 COM1 端口设置为默认属性值,单击【确定】按钮,见图 1.9。

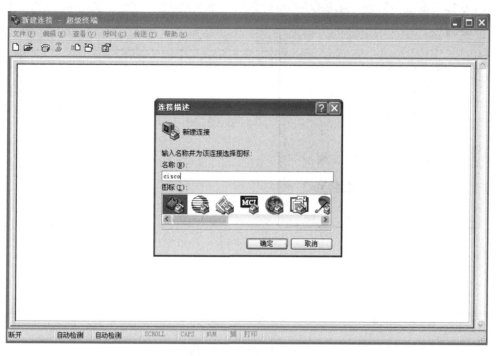

图 1.7 输入新建连接名称

图 1.8 设置连接

IOS 配置基础

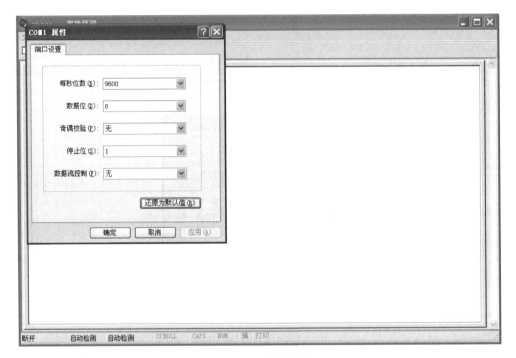

图 1.9　还原为默认值

（5）打开路由器电源，显示路由器启动信息。

```
System Bootstrap, Version 12.1(3r)T2, RELEASE SOFTWARE (fc1)
Copyright (c) 2000 by cisco Systems, Inc.
PT 1001 (PTSC2005) processor (revision 0x200) with 60416K/5120K bytes of memory

Self decompressing the image :
################################################################ [OK]

            Restricted Rights Legend

Use, duplication, or disclosure by the Government is
subject to restrictions as set forth in subparagraph
(c) of the Commercial Computer Software - Restricted
Rights clause at FAR sec. 52.227-19 and subparagraph
(c) (1) (ii) of the Rights in Technical Data and Computer
Software clause at DFARS sec. 252.227-7013.

              cisco Systems, Inc.
              170 West Tasman Drive
              San Jose, California 95134-1706

Cisco Internetwork Operating System Software
IOS (tm) PT1000 Software (PT1000-I-M), Version 12.2(28), RELEASE SOFTWARE (fc5)
Technical Support: http://www.cisco.com/techsupport
Copyright (c) 1986-2005 by cisco Systems, Inc.
```

```
Compiled Wed 27 - Apr - 04 19:01 by miwang
PT 1001 (PTSC2005) processor (revision 0x200) with 60416K/5120K bytes of memory
.
Processor board ID PT0123 (0123)
PT2005 processor: part number 0, mask 01
Bridging software.
X.25 software, Version 3.0.0.
4 FastEthernet/IEEE 802.3 interface(s)
2 Low - speed serial(sync/async) network interface(s)
32K bytes of non - volatile configuration memory.
63488K bytes of ATA CompactFlash (Read/Write)

          --- System Configuration Dialog ---

Continue with configuration dialog? [yes/no]:
```

//以上提示是否进入配置对话模式？回答"n"结束该模式。

```
Router > enable
```

//进入特权模式。

```
Router # configure terminal
```

//进入全局配置模式。

```
Enter configuration commands, one per line. End with CNTL/Z.
Router(config) # hostname router0
```

//为路由器重新命名。

```
Router0(config) # interface s2/0
```

//进入接口配置模式。

```
Router0(config - if) # exit
Router0(config) # line vty 0 4
```

//进入到虚拟终端线路配置模式。

```
Router0(config - line) # exit
```

//退出到上层模式。

```
Router0(config) # end
```

//退出到特权模式。

```
Router0 # show?
   aaa              Show AAA values
   access - lists   List access lists
   arp              Arp table
   cdp              CDP information
   class - map      Show QoS Class Map
   clock            Display the system clock
   controllers      Interface controllers status
   crypto           Encryption module
 -- More --
```

【提示】

(1) 当不知道完成某功能该使用什么命令的时候,可以在提示符下输入一个"?"查找想要的命令,这样 IOS 会列出该模式下所有的命令以供查找。

(2) 如果忘记了一个命令的拼写方法,只要给出该命令的前几个字母并且紧跟一个"?",就可以列出所有以这几个字母开始的命令,以供查找。

(3) 如果忘记了一个命令语句的句法,可以使用在该命令后面空一个格,再输入"?"的方法查询。

```
Router0#show interfaces|include f0/0
FastEthernet0/0 is administratively down, line protocol is down (disabled)
  Hardware is Lance, address is 0040.0b04.c1b9 (bia 0040.0b04.c1b9)
  MTU 1500 bytes, BW 100000 Kbit, DLY 100 usec,
     reliability 255/255, txload 1/255, rxload 1/255
  Encapsulation ARPA, loopback not set
  ARP type: ARPA, ARP Timeout 04:00:00,
  Last input 00:00:08, output 00:00:05, output hang never
  Last clearing of "show interface" counters never
  Input queue: 0/75/0 (size/max/drops); Total output drops: 0
  Queueing strategy: fifo
  Output queue :0/40 (size/max)
 --More--
```

【提示】

(1) 路由器接收一个命令时,并不需要将整词输入,只要当前输入的字母可以唯一表示某一命令即可,例如"f0/0"即表示"FastEthernet0/0"。

(2) 当使用者需要对大量的输出进行分类或者使用者想排除不需要见到的输出时,使用者可使用 show 和 more 命令搜索和过滤输出,使用这项功能,输入 show 和 more 命令,其后跟随管道字符"|",关键字 begin、include 或 exclude。

第 2 章　接口与管理配置

2.1　基 础 知 识

2.1.1　接口配置

Cisco 路由器支持两种类型的接口：物理接口和逻辑接口。在一个设备上的物理接口类型依赖于它自身的接口处理器或接口适配器。Cisco 路由器支持的逻辑接口包括 NULL 空接口、Loopback 环回接口、子接口、Tunnel 隧道接口和 Dialer 拨号接口。

为了配置一个接口，无论是物理接口还是逻辑接口，使用的配置命令和基本的配置步骤如下：

首先，在特权模式下使用 configure terminal 命令进入到全局配置模式。

然后，在全局配置模式下使用 interface 命令进入到指定接口类型和编号接口的接口配置模式。

最后，在指定接口的接口配置模式下，使用该模式下支持的接口配置命令，定义该接口的特征信息，例如封装的协议、地址信息、功能参数等，为了使配置立即生效，一般需要重新激活该接口。

2.1.2　系统管理

(1) Cisco 设备有 4 种线路：控制台线路、辅助线路、异步线路和虚拟终端线路。不同路由器有不同线路类型数。

(2) IOS 拥有 5 大口令，分别是控制台端口口令、辅助端口口令、虚拟终端端口口令、Enable password 口令、Enable secret 口令。

2.1.3　文件管理

1. 配置文件管理

使用者可将由某个源位置所指定的配置文件复制到目的地位置所指定处，这些配置文件的相关存放位置和 copy 命令的使用如图 2.1 所示。

2. 映像文件管理

复制映像的方法与复制配置文件的方法基本相同。

3. 启动文件指定

1) 指定启动配置文件

网络服务器将尝试从远端主机加载两个配置文件。第一个是网络配置文件，含有应用

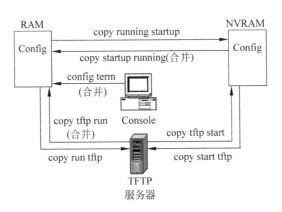

图 2.1 配置文件位置及复制

到一个网络上所有网络服务器的命令。第二个是主机配置文件,含有应用到特定的网络服务器的命令。

2) 指定启动系统映像

路由器默认在闪存中寻找 IOS 系统映像并启动。路由器也可以从下列的位置寻找系统映像:闪存、TFTP 服务器和 ROM。

2.1.4 故障处理

1. 用 show 命令显示系统信息

显示路由器信息可用于确定资源利用率和解决网络问题。

2. 测试网络连接

3. Debug 操作

Debug 特权 EXEC 命令的输出提供有关各种网际互联事件的诊断信息,这些事件一般涉及协议状态和网络行为。

4. 系统日志消息

IOS 支持系统日志(Syslog)协议将日志消息发送到指定目的地。Syslog 是一种通过网络传送事件信息的传输机制。

2.1.5 CDP 介绍

CDP(Cisco Discovery Protocol)是 Cisco 专有的协议,是使 Cisco 网络设备能够发现相邻的、直连的其他 Cisco 设备的协议。CDP 是数据链路层的协议,因此使用不同的网络层协议的 Cisco 设备也可以获得对方的信息。CDP 默认是启动的。

2.2 实验1 路由器基本配置

2.2.1 实验拓扑

一台 PC 通过 RS-232 串口使用控制台线连接一台路由器的 Console 端口,见图 2.2。

2.2.2 实验内容

通过控制台电缆,利用超级终端软件对路由器进行手工初始配置。

(1) 设置路由器名称、密码(非加密形式与加密形式)。

图 2.2 控制台方式连接路由器

(2) 开启控制台与虚拟线路初始化,并设置密码。
(3) 保存配置。
(4) 重新启动路由器,验证配置。

2.2.3 实验步骤

(1) 进入特权模式和配置模式。

```
Press RETURN to get started!

Router>enable
Router#config terminal
```

(2) 设置主机名。

```
Router(config)#hostname <hostname>
```

(3) 设置非加密口令 aaa。

```
Router(config)#enable password aaa
```

(4) 设置加密口令 bbb。

```
Router(config)#enable secret bbb
```

(5) 进入控制台初始化并设置密码。

```
Router(config)#line console 0
Router(config-line)#password ccc
```

(6) 进入虚拟终端并设置登录密码。

```
Router(config-line)#line vty 0 4
Router(config-line)#login
Router(config-line)#password ddd
```

【提示】 以上是进入路由器的 vty 虚拟终端下,"vty 0 4"表示 vty 0~vty 4,共 5 个虚拟终端。

(7) 返回退出。

```
Router(config)# end
Router# exit
```

【提示】 输入"end"或者按 Ctrl+Z 键为返回到特权模式;"exit"表示退出当前模式。

(8) 保存配置信息。

```
Router# copy running-config startup-config
Destination filename [startup-config]?
Building configuration...
[OK]
Router#
```

【提示】 只有保存运行配置,才能当机器重新启动后,运行配置中配置还能继续有效。

(9) 查看配置信息。

```
Router# show running-config
Building configuration...

Current configuration : 699 bytes
!
version 12.2
no service timestamps log datetime msec
no service timestamps debug datetime msec
no service password-encryption
!
hostname router
!
!
!
enable secret 5 $1$mERr$MBmafWD9.kwwwgArTAUad0
enable password aaa
!
!
 -- More --
```

【提示】 按【回车】键,再显示一行信息,按【空格】键,再显示其余信息,按 Esc 键,结束显示信息。

(10) 验证配置信息。

```
!
!
line con 0
  password ccc
line vty 0 4
  password ddd
```

```
    login
!
!
```

2.3 实验2 路由器高级配置命令

2.3.1 实验拓扑

一台 PC 通过 RS-232 串口使用控制台线连接一台路由器的 Console 端口，使用交叉线连接网口 Fa0/0，见图 2.3。

设备名	IP 地址	端口
Router	192.168.0.1	Fa0/0
PC	192.168.0.2	Fa

2.3.2 实验内容

通过控制台电缆，练习常用路由器高级配置命令的用法。

(1) 学习设置路由器端口 IP 地址、子网掩码等参数。

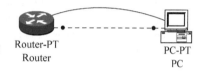

图 2.3 Telnet 方式连接路由器

(2) 查看并掌握路由器常用命令。
(3) 利用 Telnet 方式访问路由器。

2.3.3 实验步骤

(1) 使用命令 show interfaces 查看设备各端口情况。

```
Router#show interfaces
FastEthernet0/0 is up, line protocol is up (connected)
  Hardware is Lance, address is 0090.215c.5e37 (bia 0090.215c.5e37)
  Internet address is 192.168.0.1/24
  MTU 1500 bytes, BW 100000 Kbit, DLY 100 usec,
     reliability 255/255, txload 1/255, rxload 1/255
  Encapsulation ARPA, loopback not set
  ARP type: ARPA, ARP Timeout 04:00:00,
  Last input 00:00:08, output 00:00:05, output hang never
Last clearing of "show interface" counters never
```

【提示】 通过返回信息，可以获知该路由器具有 fa0/0（简写）端口，以及端口的一些参数情况。

(2) 设置端口 fa0/0 的 IP 地址。

```
Router#conf t
Enter configuration commands, one per line. End with CNTL/Z.
Router(config)#interface fa0/0
Router(config-if)#ip address 192.168.0.1 255.255.255.0
Router(config-if)#no shut
Router(config)#end
```

【提示】 命令行输入不区分大小写,当输入命令部分字母具有唯一性时可以替代该命令,或者按 Tab 键补齐。

(3) 输入"?"查看当前模式路由器可用命令。

```
Router#?
Exec commands:
  <1-99>      Session number to resume
  auto        Exec level Automation
  clear       Reset functions
  clock       Manage the system clock
  configure   Enter configuration mode
  connect     Open a terminal connection
  copy        Copy from one file to another
 -- More --
```

(4) 通过 Telnet 访问路由器。

在计算机上配置网卡的 IP 地址为 192.168.0.2/255.255.255.0,并打开 DOS 命令行窗口。

首先测试计算机和路由器的 IP 连通性,再进行 Telnet 远程登录。

```
Packet Tracer PC Command Line 1.0
PC>ping 192.168.0.1

Pinging 192.168.0.1 with 32 bytes of data:

Reply from 192.168.0.1: bytes=32 time=7ms TTL=255
Reply from 192.168.0.1: bytes=32 time=4ms TTL=255
Reply from 192.168.0.1: bytes=32 time=2ms TTL=255
Reply from 192.168.0.1: bytes=32 time=7ms TTL=255

Ping statistics for 192.168.0.1:
    Packets: Sent = 4, Received = 4, Lost = 0 (0% loss),
Approximate round trip times in milli-seconds:
    Minimum = 2ms, Maximum = 7ms, Average = 5ms
```

【提示】 以上提示表示计算机与路由器已经连通。

```
PC>telnet 192.168.0.1
Trying 192.168.0.1 ...Open
```

```
User Access Verification

Password:
Router > enable
Password:
Router#
```

【提示】 上述第一个密码为2.2节设置vty链路密码ddd,第二个密码为特权模式加密密码bbb,出现以上信息表示已经远程登录路由器。

2.4 实验3 路由器配置文件备份

2.4.1 实验拓扑

一台PC通过RS-232串口使用控制台线连接一台路由器的Console端口,使用交叉线连接网口Fa0/0,同时加入一台服务器Server使用交叉线连接路由器网口Fa1/0,见图2.4。

设备名	IP 地址	端口
Router	192.168.0.1	Fa0/0
Router	172.16.1.1	Fa1/0
PC	192.168.0.2	Fa
Server	172.16.1.2	Fa

2.4.2 实验内容

熟悉TFTP服务器的使用,学习备份与载入配置文件和操作系统。

(1) 创建TFTP服务器并建立连接。
(2) 掌握各种方式保存配置文件的方法。
(3) 掌握备份路由器IOS的方法。

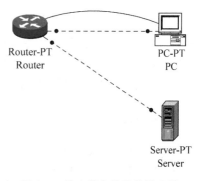

图2.4 路由器文件备份服务器

2.4.3 实验步骤

(1) 配置TFTP服务器的连接关系。

```
Router#conf t
Enter configuration commands, one per line. End with CNTL/Z.
Router(config)#int fa1/0
Router(config-if)#ip add 172.16.1.1 255.255.255.0
Router(config-if)#no shutdown

%LINK-5-CHANGED: Interface FastEthernet1/0, changed state to up
```

(2) 配置 server 的 IP 地址,确认连接可到达。

```
Router#ping 172.16.1.2

Type escape sequence to abort.
Sending 5, 100 - byte ICMP Echos to 172.16.1.2, timeout is 2 seconds:
.!!!!
Success rate is 80 percent (4/5), round - trip min/avg/max = 1/4/10 ms
```

(3) 保存配置文件到 nvram。

```
Router#copy running - config startup - config
Destination filename [startup - config]?
Building configuration...
[OK]
```

(4) 备份配置到 TFTP 服务器。

```
Router#copy running - config tftp:
Address or name of remote host []? 172.16.1.2
Destination filename [Router - confg]? r - back

Writing running - config...!!
[OK - 729 bytes]

729 bytes copied in 0.007 secs (104000 bytes/sec)
```

【提示】 除了使用 TFTP 服务器备份配置文件,也可以简单地在终端窗口中,执行"showrunning-config"命令显示当前的配置,在终端窗口中复制全部配置,粘贴到某文本文件中备份,备份文件保存在服务器根目录下。

(5) 备份配置 IOS 到 TFTP 服务器。

```
Router#show flash

System flash directory:
File Length   Name/status
  3  5571584  pt1000 - i - mz.122 - 28.bin
  4  729      running - config
  2  28282    sigdef - category.xml
  1  227537   sigdef - default.xml
[5828132 bytes used, 58188252 available, 64016384 total]
63488K bytes of processor board System flash (Read/Write)

Router#copy flash: pt1000 - i - mz.122 - 28.bin tftp:
Address or name of remote host []? 172.16.1.2
Destination filename [pt1000 - i - mz.122 - 28.bin]?
!!
```

【提示】 默认时和源文件名是一样的,不建议修改文件名,因为IOS文件名包含IOS的版本、特征等信息。

2.5 实验4 路由器的密码设置、保存与破解方法

2.5.1 实验拓扑

一台 PC 通过 RS-232 串口使用控制台线连接一台路由器的 Console 端口,使用交叉线连接网口 Fa0/0,见图 2.5。

设备名	IP 地址	端口
Router	192.168.0.1	Fa0/0
PC	192.168.0.2	Fa

2.5.2 实验内容

(1) 掌握路由器的密码设置与保存方法。
(2) 熟悉路由器的密码恢复步骤。

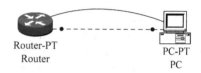

图 2.5 Telnet 方式连接路由器

2.5.3 实验步骤

(1) 在路由器上配置密码。

```
Router>enable
Router#configure terminal
Enter configuration commands, one per line. End with CNTL/Z.
Router(config)#hostname R1
//更改主机名称.
R1(config)#enable password 123456abc
R1(config)#ip host pc 192.168.0.2
R1(config)#end
R1#ping pc
Type escape sequence to abort.
Sending 5, 100 - byte ICMP Echos to 192.168.0.2, timeout is 2 seconds:
.!!!!
Success rate is 80 percent (4/5), round - trip min/avg/max = 2/4/6 ms
```

//建立 IP 地址与主机映射关系

(2) 密码及配置文件保存方法。

```
R1#copy running - config startup - config
Destination filename [startup - config]?
Building configuration...
[OK]
```

(3) 路由器密码恢复。

关闭路由器电源并重新开机,当控制台出现启动过程,赶快按 Ctrl+Break 键中断路由器的启动过程,进入 rommon 模式,如下:

```
System Bootstrap, Version 12.4(1r) [hqluong 1r], RELEASE SOFTWARE (fc1)
Copyright (c) 2005 by cisco Systems, Inc.
Initializing memory for ECC
c2821 processor with 262144 Kbytes of main memory
Main memory is configured to 64 bit mode with ECC enabled
Readonly ROMMON initialized
rommon 1 > confreg 0x2142
```

//改变配置寄存器的值为 0x2142,这会使得路由器开机时不读取 NVRAM 中的配置文件。

```
rommon 2 > i
```

//重启路由器。路由器重启后会直接进入到 setup 配置模式,按 Ctrl+C 键或者回答"n",退出 setup 模式。

```
Router > enable
Router # copy startup - config running - config
Destination filename [running - config]?
661 bytes copied in 0.625 secs
```

【提示】 注意复制文件前提示符的变化,把配置文件从 NVRAM 中复制到 RAM 中,在此基础上修改密码。

```
R2 # configure terminal
R2(config) # enable password 123456
```

//以上把密码改为自己的密码,如果还配置别的密码则一起把它们修改了。

```
R2(config) # config - register 0x2102
```

//以上把寄存器的值恢复为正常值 0x2102。

```
R2(config) # exit
R2 # copy running - config startup - config
Destination filename [startup - config]?
Building configuration...
[OK]
R2 # reload
```

//以上是保存配置,重启路由器,检查路由器是否正常

【提示】 在保存配置前,还需要把路由器的各个接口一一打开。

2.6 实验5 CDP

2.6.1 实验拓扑

一台 PC 通过 RS-232 串口使用控制台线连接一台路由器的 Console 端口,使用交叉线

连接网口 Fa0/0,见图 2.6。

设备名	IP 地址	端口
Router0	192.168.0.1	Fa0/0
Router0	16.16.1.1	S2/0
Router1(DCE)	172.16.0.1	Fa0/0
Router1(DCE)	16.16.1.2	S2/0

2.6.2 实验内容

(1) 熟悉 CDP 原理与相关指令。
(2) 查找 CDP 邻居信息。

2.6.3 实验步骤

(1) 打开相对应接口,配置 IP 地址并设置传输率。

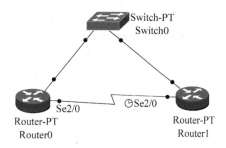

图 2.6 CDP 验证

```
Router0(config)# int fa0/0
Router0(config-if)# no shut
Router0(config-if)# int s2/0
Router0(config-if)# no shut
Router0(config-if)# end

Router1(config)# int fa0/0
Router1(config-if)# no shut
Router1(config-if)# int s2/0
Router1(config-if)# clock rate 64000
Router1(config-if)# no shut
Router1(config-if)# end
```

【提示】 以上省略 IP 地址配置过程,注意传输率设置在 DCE 设备上即 router1。
(2) 查看 CDP 配置。

```
Router0# show cdp
Global CDP information:
    Sending CDP packets every 60 seconds
    Sending a holdtime value of 180 seconds
Sending CDPv2 advertisements is enabled

Router0# show cdp interface
FastEthernet0/0 is up, line protocol is up
  Sending CDP packets every 60 seconds
  Holdtime is 180 seconds
FastEthernet1/0 is administratively down, line protocol is down
  Sending CDP packets every 60 seconds
  Holdtime is 180 seconds
```

```
Serial2/0 is up, line protocol is up
   Sending CDP packets every 60 seconds
   Holdtime is 180 seconds
Serial3/0 is administratively down, line protocol is down
   Sending CDP packets every 60 seconds
   Holdtime is 180 seconds
FastEthernet4/0 is administratively down, line protocol is down
   Sending CDP packets every 60 seconds
   Holdtime is 180 seconds
FastEthernet5/0 is administratively down, line protocol is down
   Sending CDP packets every 60 seconds
Holdtime is 180 seconds
```

//以上显示在哪些接口运行 CDP。

(3) 查看 CDP 邻居信息。

```
Router0 # show cdp neighbors
Capability Codes: R - Router, T - Trans Bridge, B - Source Route Bridge
                 S - Switch, H - Host, I - IGMP, r - Repeater, P - Phone
Device ID    Local Intrfce   Holdtme   Capability   Platform   Port ID
Switch       Fas 0/0         174       S            PT3000     Fas 0/1
router1      Ser 2/0         171       R            PT1000     Ser 2/0
```

//以上显示 Router0 路由器有两个邻居：Router1 和 Switch。"Device ID"表示邻居的主机名；"Local Intrfce"表明 Router0 通过该接口和邻居连接，注意是 Router0 上的接口；"Holdtme"指收到邻居发送的 CDP 消息的时间，采用倒计时；"Capability"表明邻居是什么设备，第一、二行 Capability Codes 对各符号进行了说明；"Platform"指明了邻居的硬件型号；"Port ID"指明了 Router0 是连接对方的哪个接口上。

```
Router0 # show cdp entry router1

Device ID: router1
Entry address(es):
   IP address : 16.16.1.2
Platform: cisco PT1000, Capabilities: Router
Interface: Serial2/0, Port ID (outgoing port): Serial2/0
Holdtime: 141

Version :
Cisco Internetwork Operating System Software
IOS (tm) PT1000 Software (PT1000 - I - M), Version 12.2(28), RELEASE SOFTWARE (fc5)
Technical Support: http://www.cisco.com/techsupport
Copyright (c) 1986 - 2005 by cisco Systems, Inc.
Compiled Wed 27 - Apr - 04 19:01 by miwang

advertisement version: 2
Duplex: full
```

//以上是显示邻居 Router1 的详细信息。

```
Router0 # clear cdp table
```

//清除 CDP 表。

```
Router0(config)#int f0/0
Router0(config-if)#no cdp enable
```

//端口关闭 CDP。

```
Router0(config-if)#exit
Router0(config)#no cdp run
```

//整个路由器关闭 CDP。

第 3 章　IP 特性配置

3.1　基 础 知 识

3.1.1　地址解析方法

1. 静态映射 IP 地址到 MAC 地址

ARP(Address Resolution Protocol,地址解析协议)是一个位于 TCP/IP 协议栈中的低层协议,负责将某个 IP 地址解析成对应的 MAC 地址。

当一个基于 TCP/IP 的应用程序需要从一台主机发送数据给另一台主机时,它把信息分割并封装成包,附上目的主机的 IP 地址。然后,寻找 IP 地址到实际 MAC 地址的映射,这需要发送 ARP 广播消息。当 ARP 找到了目的主机 MAC 地址后,就可以形成待发送帧的完整以太网帧头。最后,协议栈将 IP 包封装到以太网帧中进行传送。

2. 映射主机域名到 IP 地址

DNS 是 Domain Name System(域名系统)的缩写,该系统用于命名组织到域层次结构中的计算机和网络服务。在 Internet 上域名与 IP 地址之间是一对一(或者一对多)的,域名虽然便于人们记忆,但机器之间只能互相认识 IP 地址,它们之间的转换工作称为域名解析,域名解析需要由专门的域名解析服务器来完成,DNS 就是进行域名解析的服务器。DNS 命名用于 Internet 等 TCP/IP 网络中,通过用户友好的名称查找计算机和服务。当用户在应用程序中输入 DNS 名称时,DNS 服务可以将此名称解析为与之相关的其他信息,如 IP 地址。

3.1.2　广播包处理

广播是以某个特定物理网络上的所有主机为目的地的数据包。主机通过特殊地址识别广播。广播被几个重要的 Internet 协议使用。因为广播包有使网络超载的潜在可能,所以控制广播包对一个 IP 网络的正常运行是必要的。

IOS 支持两种类型的广播:定向广播和泛洪广播(也称为全向广播)。

3.1.3　自治系统

AS(自治系统)是在单一技术管理下,采用同一种内部网关协议和统一度量值在 AS 内转发数据包,并采用一种外部网关协议将数据包转发到其他 AS 的一组路由器。AS 是网络的集合,或更确切地说,是连接这些网络的路由器的集合,这些路由器位于同一管理机构管理之下并共享共同的路由策略。

3.1.4 路由技术

所谓路由,是被用来把来自一台设备的数据包穿过网络发送到位于另一个网段的设备上的路径信息。它具体表现为路由表里的条目。

路由技术就是使路由器学习到路由,对路由进行控制,并且维护这些路由的完整,无差错的技术。要想使路由有效地工作,必须具备以下条件:

(1) 要知道目的地址。假如不知道数据包的目的地址,就没办法为数据包路由了。

(2) 有可以学到路由的资源。这包括两个方面:路由器要么从相邻的路由器那里学到路由,要么由网络管理员手动地配置路由。但是,有一种路由是不用学就可以得到的,那就是直接连接在该路由器的接口上的网段。

(3) 有可能到达目的网络的路径。只是有了可以学到路由的资源还是不行,这些资源里没有人知道到达目的地的路径,还是一样不能路由。

(4) 在众多可能到达目的 IP 地址的路径中有最佳的路由。一般情况下,可能会有多条路径到达目的网段,还需要在它们中间选择最佳的路径作为路由。

(5) 管理和维护路由信息。如果出现了错误的路由,后果是很严重的,数据包会被发往到那个错误的位置,网络就不通畅,甚至完全地瘫痪掉。

以上这些方面,都是实施路由技术所要考虑的,所有路由技术不只是使用什么路由协议的问题,它是一整套关于如何实施路由的策略。

3.1.5 路由表

在计算机网络中,路由表或称路由择域信息库(RIB)是一个存储在路由器或者联网计算机中的电子表格(文件)或者数据库。路由表存储着指向特定网络地址的路径。路由表中含有网络周边的拓扑信息。路由表建立的主要目标是为了实现路由协议和静态路由选择。

3.1.6 管辖距离

管辖距离(Administrative Distance,AD)是在路由上附加的一个度量,用来描述路由的可信度。不同的路由协议使用的逻辑和度量都是不同的,当同时使用多个路由协议时,路由器必须知道哪一个给出的是最准确的信息。Cisco 解决这个问题是通过给每个路由协议指定一个管辖距离,在查看了管辖距离之后才看度量。

3.1.7 度量值

所谓度量值(Metric)就是路由协议根据自己的路由算法计算出来的一条路径的优先级。当有多条路径到达同一个目的地时,度量值最小的路径是最佳的路径,应该进入路由表。

各种路由协议的度量值各不相同。例如,RIP 是用路径上经过路由器的数量(也就是跳数)作为度量值,OSPF 协议则是用路径的带宽来计算度量值的。

3.1.8 路由更新

拓扑结构的变化与网络上每台路由器的路由表作出的相应改变之间所需的时间称为收

敛时间。收敛是指为使所有的路由表具有一致的信息并且处于稳定状态所作出的动作。使路由协议具有短的收敛时间是十分必要的，因为在路由器计算新的路由时可能会产生路由中断或路由环。

路由更新可以包含路由器的整个路由表，或者仅包含变化的那部分。这些通信对于路由器及时接收到关于路由环境的可信的完整信息，保持路由表的准确性，以及允许选择最优路由是必不可少的。根据所使用的路由协议，路由更新可以定期发出，或者可以由拓扑结构的变动而触发。

3.1.9 路由查找

一般地，路由器查找路由的顺序是直连网络、静态路由、动态路由，如果以上路由表中都没有合适的路由，则通过默认路由将包传输出去。综合使用动态路由、静态路由和默认路由，以保证路由的冗余。

初始时路由器使用直连网络或子网的路由，然后扫描路由表中匹配项目的操作步骤如下：

（1）首先使用子网掩码来确定数据包的网络地址，并且对路由表进行扫描。如果在路由表中有多条与目的地匹配的表项，则最长匹配查找（即用尽量长的子网掩码位）与该网络地址相匹配的路由，并将数据包发送给该路由表项中设定的下一跳地址。

（2）如果不存在这样的一个路由表项，就寻找一个默认路由，然后就将数据包发送给该路由表项中设定的下一跳地址。

（3）如果不存在这样的一个默认路由，IP 就将一个 ICMP"目的不可达"的消息发送给数据包的源系统。

3.1.10 静态路由和动态路由

静态路由是由网络管理员手动配置在路由器的路由表里的路由。静态路由的基本思想是，如果想要路由器知道某个网络，就手工输入这些路径。静态路由十分容易理解也十分容易配置，至少在一个小型网络中是如此，而且无疑是最简单的路由方法。

动态路由是网络中的路由器之间相互通信，传递路由信息，利用收到的路由信息更新路由器表的过程。它能实时地适应网络结构的变化。如果路由更新信息表明发生了网络变化，路由选择软件就会重新计算路由，并发出新的路由更新信息。这些信息通过各个网络，引起各路由器重新启动其路由算法，并更新各自的路由表以动态地反映网络拓扑变化。动态路由适用于网络规模大、网络拓扑复杂的网络。当然，各种动态路由协议会不同程度地占用网络带宽和 CPU 资源。

静态路由和动态路由有各自的特点和适用范围，因此在网络中动态路由通常作为静态路由的补充。当一个分组在路由器中进行寻径时，路由器首先查找静态路由，如果查到则根据相应的静态路由转发分组；否则再查找动态路由。

3.1.11 默认路由

默认路由（default route），是对 IP 数据包中的目的地址找不到存在的其他路由时，路由器所选择的路由。目的地不在路由器的路由表里的所有数据包都会使用默认路由。这条路由一般会连去另一个路由器,而这个路由器也同样处理数据包;如果知道应该怎么路由这个数据包,则数据包会被转发到已知的路由;否则,数据包会被转发到默认路由,从而到达另一个路由器。每次转发,路由都增加了一跳的距离。

3.1.12 汇总路由

路由汇总又称路由汇聚。路由汇总的含义是把一组路由汇总为一个单个的路由广播。路由汇总的最终结果和最明显的好处是缩小网络上的路由表的尺寸。这样将减少与每一个路由跳有关的延迟,因为由于减少了路由登录项数量,查询路由表的平均时间将加快。由于路由登录项广播的数量减少,路由协议的开销也将显著减少。随着整个网络(以及子网的数量)的扩大,路由汇总将变得更加重要。

除了缩小路由表的尺寸之外,路由汇总还能通过在网络连接断开之后限制路由通信的传播来提高网络的稳定性。如果一台路由器仅向下一个下游的路由器发送汇聚的路由,那么,它就不会广播与汇聚的范围内包含的具体子网有关的变化。

3.1.13 DHCP 概述

动态主机控制协议(DHCP)能够自动分配重用 IP 地址到 DHCP 客户。Cisco IOS DHCP 服务器是完整的 DHCP 服务器,用于分配和管理来自在路由器上指定地址池的 IP 地址给 DHCP 客户。如果 Cisco IOS DHCP 服务器不能根据自己的数据库满足 DHCP 请求,它能够转发请求到一个或者多个辅助由网络管理员定义的 DHCP 服务器。

图 3.1 显示了当 DHCP 客户从 DHCP 服务器请求一个 IP 地址发生的基本过程。

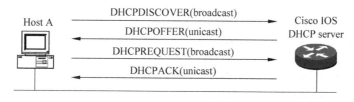

图 3.1 DHCP 工作基本步骤

3.2 实验 1 路由器的背对背连接

3.2.1 实验拓扑

两台路由器通过串口线背对背连接,定义 Router0 为 DTE 设备,Router1 为 DCE 设备,传输率为 64 000k,见图 3.2。

设备名	IP 地址	端口
Router0	192.168.0.1	S2/0
Router1	192.168.0.2	S2/0

图 3.2 路由器背对背连接

3.2.2 实验内容

通过对路由器 router0 和路由器 router1 的配置,练习在串行线路上的配置,使路由器可以通过串行线路通信。

(1) 掌握路由器的 IP 地址和传输率设置方法。

(2) 进行连通测试。

3.2.3 实验步骤

(1) 设置 IP 地址和传输率。

```
Router0 > ena
Router0 # conf t
Enter configuration commands, one per line. End with CNTL/Z.
Router0(config) # int s2/0
Router0(config - if) # ip add 192.168.0.1 255.255.255.0
Router0(config - if) # no shut

% LINK - 5 - CHANGED: Interface Serial2/0, changed state to down

Router1 > ena
Router1 # conf t
Enter configuration commands, one per line. End with CNTL/Z.
Router1(config) # int s2/0
Router1(config - if) # ip add 192.168.0.2 255.255.255.0
Router1(config - if) # clock rate 64000
Router1(config - if) # no shut

% LINK - 5 - CHANGED: Interface Serial2/0, changed state to up
```

【提示】 串行通信需要设置 DTE 设备和 DCE 设备,由 DCE 设备提供传输率,传输率设置可以根据提示选择相应数值。

(2) 连通测试。

```
Router # ping 192.168.0.2

Type escape sequence to abort.
Sending 5, 100 - byte ICMP Echos to 192.168.0.2, timeout is 2 seconds:
!!!!!
Success rate is 100 percent (5/5), round - trip min/avg/max = 2/4/7 ms
```

3.3 实验2 静态路由协议

3.3.1 实验拓扑

两台路由器直连,分别管控两个子网,子网由 PC1 和 PC2 模拟构成,见图 3.3。

设备名	IP 地址	端口
Router0	192.168.0.1	Fa0/0
Router0	16.16.1.1	S2/0
Router1(DCE)	172.16.0.1	Fa0/0
Router1(DCE)	16.16.1.2	S2/0
PC1	192.168.0.2	Fa0
PC2	172.16.0.2	Fa0

3.3.2 实验内容

掌握路由器静态路由配置方法。
(1) 掌握静态路由配置命令使用方法。
(2) 查看路由表项。
(3) 正确配置静态路由实现连通。

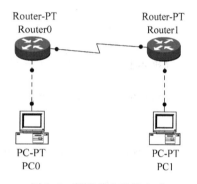

图 3.3 静态路由连接方式

3.3.3 实验步骤

(1) 路由器上配置 IP 地址、传输率,实现相邻设备连通。

```
Router0 # conf t
Enter configuration commands, one per line. End with CNTL/Z.
Router0(config) # int s2/0
Router0(config - if) # ip add 16.16.1.1 255.255.255.0
Router0(config - if) # no shut
Router0(config - if) # int f0/0
Router0(config - if) # ip add 192.168.0.1 255.255.255.0
Router0(config - if) # no shut
Router0(config - if) # end
```

```
Router1 # conf t
Enter configuration commands, one per line. End with CNTL/Z.
Router1(config) # int s2/0
Router1(config - if) # ip add 16.16.1.2 255.255.255.0
Router1(config - if) # clock rate 64000
Router1(config - if) # no shut
Router1(config - if) # int f0/0
Router1(config - if) # ip add 172.16.0.1 255.255.255.0
```

```
Router1(config-if)#no shut
Router1(config-if)#end


Router1#ping 16.16.1.1

Type escape sequence to abort.
Sending 5, 100-byte ICMP Echos to 16.16.1.1, timeout is 2 seconds:
!!!!!
Success rate is 100 percent (5/5), round-trip min/avg/max = 4/11/36 ms
```

(2) 配置静态路由。

```
Router0(config)#ip route 172.16.0.0 255.255.255.0 serial 2/0
Router0(config)#end
```

//下一跳为接口形式,s2/0 是点对点的链路,注意应该是 router0 上的 s2/0 接口。

```
Router1(config)#ip route 192.168.0.0 255.255.255.0 s2/0
Router1(config)#end
```

【提示】 也可以配置下一跳为 IP 地址形式,例如:

```
Router0(config)#ip route 172.16.0.0 255.255.255.0 16.16.1.2
```

其中,"16.16.1.2"为 router1 的 IP 地址。

(3) 测试连通性,并查看路由表。

```
PC1>ping 192.168.0.2

Pinging 192.168.0.2 with 32 bytes of data:

Reply from 192.168.0.2: bytes=32 time=18ms TTL=126
Reply from 192.168.0.2: bytes=32 time=15ms TTL=126
Reply from 192.168.0.2: bytes=32 time=8ms TTL=126
Reply from 192.168.0.2: bytes=32 time=33ms TTL=126

Ping statistics for 192.168.0.2:
    Packets: Sent = 4, Received = 4, Lost = 0 (0% loss),
Approximate round trip times in milli-seconds:
Minimum = 8ms, Maximum = 33ms, Average = 18ms
```

//通过测试验证两子网连通成功。

```
Router0#show ip route
Codes: C - connected, S - static, I - IGRP, R - RIP, M - mobile, B - BGP
       D - EIGRP, EX - EIGRP external, O - OSPF, IA - OSPF inter area
       N1 - OSPF NSSA external type 1, N2 - OSPF NSSA external type 2
       E1 - OSPF external type 1, E2 - OSPF external type 2, E - EGP
       i - IS-IS, L1 - IS-IS level-1, L2 - IS-IS level-2, ia - IS-IS inter area
       * - candidate default, U - per-user static route, o - ODR
```

```
          P - periodic downloaded static route

Gateway of last resort is not set

     16.0.0.0/24 is subnetted, 1 subnets
C       16.16.1.0 is directly connected, Serial2/0
     172.16.0.0/24 is subnetted, 1 subnets
S       172.16.0.0 is directly connected, Serial2/0
C    192.168.0.0/24 is directly connected, FastEthernet0/0
```

//根据提示信息,可以了解路由器router0 的连通情况,两个直连网络和一个静态路由协议网络。

3.4 实验 3 默认路由

3.4.1 实验拓扑

两台路由器直连,分别管控两个子网,子网由 PC1 和 PC2 模拟构成,见图 3.4。

设备名	IP 地址	端口
Router0	192.168.0.1	Fa0/0
Router0	16.16.1.1	S2/0
Router1(DCE)	172.16.0.1	Fa0/0
Router1(DCE)	16.16.1.2	S2/0
PC1	192.168.0.2	Fa0
PC2	172.16.0.2	Fa0

3.4.2 实验内容

(1) 了解默认路由工作原理。
(2) 掌握默认路由配置过程。

3.4.3 实验步骤

该实验可以在静态路由基础上进行。
(1) 删除原有静态路由。

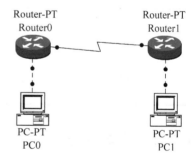

图 3.4 设置默认路由协议

```
Router0(config)# no ip route 172.16.0.0 255.255.255.0 s2/0
```

```
Router1(config)# no ip route 192.168.0.0 255.255.255.0 s2/0
```

//删除原有命令可以在原命令前加"no"。

(2) 配置默认路由。

```
Router0(config)# ip route 0.0.0.0 0.0.0.0 serial 2/0
```

```
Router1(config)# ip route 0.0.0.0 0.0.0.0 s2/0
```

(3) 连通测试。

```
Router1#ping
Protocol [ip]:
Target IP address: 192.168.0.1
Repeat count [5]:
Datagram size [100]:
Timeout in seconds [2]:
Extended commands [n]:
Sweep range of sizes [n]:
Type escape sequence to abort.
Sending 5, 100 - byte ICMP Echos to 192.168.0.1, timeout is 2 seconds:
!!!!!
Success rate is 100 percent (5/5), round - trip min/avg/max = 1/2/5 ms
```

【提示】 首先,验证默认路由与静态路由之间的区别;然后,了解 ping 命令不带参数形式的使用方法。

3.5 实验4 DHCP 基本配置

3.5.1 实验拓扑

一台路由器通过一台交换机管辖三台 PC,为其动态分配 IP 等网络参数,见图 3.5。

设备名	IP 地址	端口
Router0	192.168.0.1	Fa0/0
PC0	192.168.0.2	Fa0
PC1	192.168.0.3	Fa0
PC2	192.168.0.4	Fa0

3.5.2 实验内容

(1) 掌握 DHCP 工作原理和工作方法。
(2) DHCP 服务器基本配置。
(3) 客户端配置。

3.5.3 实验步骤

(1) 配置路由器提供 DHCP 服务。

```
Router(config)# ip dhcp pool pool1
```
//定义地址池,命名 pool1。

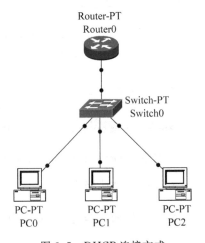

图 3.5 DHCP 连接方式

```
Router(dhcp-config)#network 192.168.0.0 255.255.255.0
Router(dhcp-config)#default-router 192.168.0.1
```

//定义默认网关,即连接端口 f0/0。

```
Router(dhcp-config)#dns-server 192.168.0.5
```

//定义 DNS 服务器。

```
Router(dhcp-config)#option 150 ip 192.168.0.6
```

//定义 TFTP 服务器。

```
Router(dhcp-config)#exit
Router(config)#ip dhcp excluded-address 192.168.0.2 192.168.0.10
Router(config)#end
```

//保留 IP 地址范围。

(2) 设置 PC 客户端,见图 3.6。

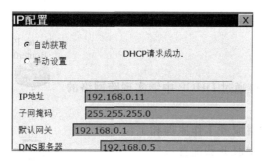

图 3.6 设置 PC 客户端

//将 IP 配置改为自动获取,可以看到自动分配 IP 地址、DNS 服务器等参数。

```
PC0 > ipconfig /all

Physical Address................: 00E0.F96D.BD5C
IP Address......................: 192.168.0.11
Subnet Mask.....................: 255.255.255.0
Default Gateway.................: 192.168.0.1
DNS Servers.....................: 192.168.0.5
```

(3) 查看路由器配置情况。

```
Router#show ip dhcp binding
IP address          Client-ID/           Lease expiration         Type
                    Hardware address
192.168.0.11        00E0.F96D.BD5C       --                       Automatic
192.168.0.12        00D0.585D.8DA2       --                       Automatic
192.168.0.13        00E0.8F77.EAE2       --                       Automatic
```

//以上表示 DHCP 服务器自动分配给客户端的 IP 地址以及对应的 MAC 地址情况。

3.6 实验 5　DHCP 中继配置

3.6.1 实验拓扑

如图 3.7 所示，Router0 担任 DHCP 服务器的角色，负责向 PC0 所在网络和 PC1 所在网络的主机动态分配 IP 地址，所以 Router0 上需要定义两个地址池，Router1 负责中继。整个网络运行 RIPv2 协议，确保网络 IP 连通性。

设备名	IP 地址	端口
Router0	192.168.0.1	Fa0/0
Router0	16.16.1.1	S2/0
Router1	172.16.0.1	Fa0/0
Router1	16.16.1.2	S2/0
PC0	——	Fa0
PC1	——	Fa0

3.6.2 实验内容

通过本实验可以掌握通过 DHCP 中继实现跨网络的 DHCP 服务。

(1) 掌握 DHCP 中继服务工作原理和工作方法。
(2) 掌握 DHCP 中继服务器配置过程。

3.6.3 实验步骤

(1) 配置路由器 Router0 的 DHCP 服务。

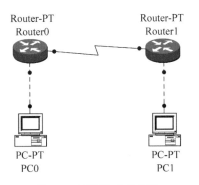

图 3.7　DHCP 中继实验

```
Router > ena
Router # conf t
Enter configuration commands, one per line. End with CNTL/Z.
Router(config) # hos router0
Router0(config) # int f0/0
Router0(config - if) # ip add 192.168.0.1 255.255.255.0
Router0(config - if) # no shut

% LINK - 5 - CHANGED: Interface FastEthernet0/0, changed state to up

% LINEPROTO - 5 - UPDOWN: Line protocol on Interface FastEthernet0/0, changed state to up

Router0(config - if) # exit
Router0(config) # router rip
Router0(config - router) # net 192.168.0.0
Router0(config - router) # net 16.16.1.0
Router0(config - router) # exit
```

//使用 RIP 路由协议,保障网络通畅,后面章节具体介绍。

```
Router0(config)# ip dhcp pool pool0
Router0(dhcp-config)# net 192.168.0.0 255.255.255.0
Router0(dhcp-config)# def 192.168.0.1
Router0(dhcp-config)# dns-server 192.168.0.3
Router0(dhcp-config)# option 150 ip 192.168.0.4
Router0(dhcp-config)# exit
Router0(config)# ip dhcp excluded-address 192.168.0.1 192.168.0.5
```

//定义 PC0 所在地址池。

```
Router0(config)# ip dhcp pool pool1
Router0(dhcp-config)# network 172.16.0.0 255.255.255.0
Router0(dhcp-config)# default-router 172.16.0.1
Router0(dhcp-config)# dns-server 192.168.0.3
Router0(dhcp-config)# option 150 ip 192.168.0.4
Router0(dhcp-config)# exit
Router0(config)# ip dhcp excluded-address 172.16.0.1 172.16.0.5
```

//定义 PC1 所在地址池。

(2) 配置路由器 Router1 的中继服务。

```
Router# conf t
Enter configuration commands, one per line. End with CNTL/Z.
Router(config)# hos router1
Router1(config)# int f0/0
Router1(config-if)# ip add 172.16.0.1 255.255.255.0
Router1(config-if)# no shut

% LINK-5-CHANGED: Interface FastEthernet0/0, changed state to up

% LINEPROTO-5-UPDOWN: Line protocol on Interface FastEthernet0/0, changed state to up

Router1(config-if)# int s2/0
Router1(config-if)# ip add 16.16.1.2 255.255.255.0
Router1(config-if)# clock rate 64000
Router1(config-if)# no shut

% LINK-5-CHANGED: Interface Serial2/0, changed state to down
Router1(config-if)# int f0/0
Router1(config-if)# ip helper-address 16.16.1.1
```

//配置中继路由器地址。

```
Router1(config-if)# no shut
Router1(config-if)# exit
Router1(config)# router rip
Router1(config-router)# net 16.16.1.0
Router1(config-router)# net 172.16.0.0
Router1(config-router)# end
```

（3）客户端测试。

```
PC0 > ipconfig /all
Physical Address...............: 0060.4720.1D1B
IP Address....................: 192.168.0.6
Subnet Mask...................: 255.255.255.0
Default Gateway...............: 192.168.0.1
DNS Servers...................: 192.168.0.3
```

```
PC1 > ipconfig /all
Physical Address...............: 0001.4373.EE79
IP Address....................: 172.16.0.6
Subnet Mask...................: 255.255.255.0
Default Gateway...............: 172.16.0.1
DNS Servers...................: 192.168.0.3
```

（4）路由器中继测试。

```
Router0 # show ip dhcp binding
IP address        Client - ID/         Lease expiration        Type
                  Hardware address
192.168.0.6       0060.4720.1D1B       --                      Automatic
172.16.0.6        0001.4373.EE79       --                      Automatic
```

```
Router1 # show ip interface
FastEthernet0/0 is up, line protocol is up (connected)
  Internet address is 172.16.0.1/24
  Broadcast address is 255.255.255.255
  Address determined by setup command
  MTU is 1500
Helper address is 16.16.1.1
……
```

//可以看到 f0/0 使用了帮助地址 16.16.1.1。

第 4 章　广域网配置

4.1　基础知识

4.1.1　DDN 概述

DDN 是利用数字信道传输数据信号的数据传输网。它的主要作用是向用户提供永久性和半永久性连接的数字数据传输信道,既可用于计算机之间的通信,也可用于传送数字化传真、数字话音、数字图像信号或其他数字化信号。

4.1.2　HDLC 概述

HDLC 使用同步串行传输,它在并不可靠的物理层链路上为两点间提供无差错通信。HDLC 定义了第 2 层帧结构,通过使用确认和窗口操作方案进行流量控制和差错控制。不管是数据帧还是控制帧,格式都相同。

标准的 HDLC 原本不支持在单一链路上传输多种协议,这是因为 HDLC 无法标识所携带的是哪一种协议。Cisco 提出了一种专有的 HDLC 版本。Cisco 的 HDLC 帧中使用了一个专有类型字段,这个字段等同于一个协议字段,使 HDLC 在同一链路中传输多种网络层协议数据成为可能。

4.1.3　PPP 概述

PPP 是点对点封装协议,它是 SLIP(串行线路 IP 协议的继承者),它提供了跨越同步和异步电路实现路由器到路由器和主机到网络的连接。PPP 是目前使用最广泛的广域网协议,这是因为它具有以下特征:

(1) 能够控制数据链路的建立。
(2) 能够对 IP 地址进行分配和管理。
(3) 支持多种网络协议的复用。
(4) 能够配置链路并对链路进行质量测试。
(5) 错误检测。

在 Cisco 路由器之间用同步专线连接时,采用 Cisco HDLC 比采用 PPP 效率高得多,但是,假如将 Cisco 路由器和非 Cisco 路由器进行同步专线连接时,不能采用 Cisco HDLC,因为它们不支持 Cisco HDLC,只能采用 PPP。

在 PPP 的点对点通信中,可以采用 PAP 或者 CHAP 身份验证方式(首选 CHAP 方式)对连接用户进行身份验证,以防非法用户的 PPP 连接。

1. PAP

PAP(Password Authentication Protocol,密码验证协议)利用两次握手的简单方法进行认证。在 PPP 链路建立完毕后,源节点不停地在链路上反复发送用户名和密码,直到验证通过。PAP 的验证中,密码在链路上是以明文传输的,而且由于是源节点控制验证重试频率和次数,因此 PAP 不能防范再生攻击和重复的尝试攻击。

2. CHAP

CHAP(Challenge Handshake Authentication Protocol,询问握手验证协议)利用三次握手周期地验证源端节点的身份。CHAP 验证过程在链路建立之后进行,而且在以后的任何时候都可以再次进行。这使得链路更为安全;CHAP 不允许连接发起方在没有收到询问消息的情况下进行验证尝试。CHAP 每次使用不同的询问消息,每个消息都是不可预测的唯一的值,CHAP 不直接传送密码,只传送一个不可预测的询问消息,以及该询问消息与密码经过 MD5 加密运算后的加密值。所以 CHAP 可以防止再生攻击,CHAP 的安全性比 PAP 要高。

4.1.4 帧中继概述

帧中继是工作在数据链路层的协议,使用的是 HDLC 的一个变种子集 LAPF(Link Access Procedure for Frame-relay)。它是面向连接的,采用包交换(packet-switch)技术。FR 采用虚电路(VC)为终端用户建立连接。有 SVC 和 PVC 两种形式。SVC 指通信前双方通过信令消息来动态建立链路;而 PVC(永久虚电路)是预设在交换机里面的。一般情况下帧中继采用的是 PVC。

1. 帧中继 DLCI 编号

帧中继技术提供面向连接的数据链路层的通信。在每对设备之间都存在一条定义好的通信链路,且该链路有一个链路识别码。这种服务通过帧中继虚电路实现,每条虚电路都用 DLCI 标识自己。

2. 帧中继 LMI 类型

LMI 是帧中继中一项重要指标,是对基本的帧中继的扩展。它是用户终端设备和帧中继交换机之间传送信令的标准,负责管理设备连接并且维护设备之间的连接状态,提供帧中继的管理机制等。LMI 分为三种类型:cisco、ansi 及 q933a。

3. 帧中继映射

路由器可以从路由选择表决定下一跳地址,但该地址必须被解析到一个帧中继 DLCI。这个解析过程是通过一个称为帧中继映射的数据结构来完成的。路由选择表用来提供出站通信流量的下一跳协议地址或 DLCI。该数据结构可在路由器中静态配置,或使用翻转 ARP 功能特性来自动建立映射。

1) 静态映射

管理员手工输入的映射就为静态映射,其命令为:

frame - relay map ip *protocol address dlci* [*broadcast*]

其中:

protocol:协议类型。

address：网络地址。
dlci：所需要交换逆向 ARP 信息的本地接口的 DLCI 号。
broadcast：表示允许在帧中继线路上传送路由广播或组播信息。

2) 动态映射

如果翻转 ARP 被禁用，为在指定接口、子接口或 DLCI 重新启用翻转 ARP，在接口配置模式下，使用 frame-relay inverse-arp 命令，为禁用翻转 ARP，使用该命令的 no 形式。

frame - relay inverse - arp [protocol] dlci
no frame - relay map protocol protocol - address

4. 帧中继子接口

当一台路由器的一个物理接口上只有一条 PVC 的时候，广播和水平分割都可以正常运作。但当一台路由器的一个物理接口存在多条 PVC 的时候，会造成广播和水平分割，对数据报的可达性造成了一定的影响。为了解决上面的问题，使路由器能够正常地转发路由更新广播，就必须在路由器连接多条 PVC 的接口上开子接口。

子接口可以分为如下两种类型。
(1) 点对点连接类型。
(2) 点对多点连接类型。

4.2 实验 1 HDLC 和 PPP 封装

4.2.1 实验拓扑

路由器 Router0 和 Router1 串口连接，验证封装方式，见图 4.1。

设备名	IP 地址	端口
Router0	16.16.1.1	S2/0
Router1	16.16.1.2	S2/0

4.2.2 实验内容

通过本实验可以掌握路由器封装形式。
(1) 了解 HDLC 封装形式。
(2) 配置 PPP 封装形式。

图 4.1 路由器串口连接方式

4.2.3 实验步骤

(1) 配置 router0 和 router1 的 IP 地址，实现链路通畅。

```
Router0(config)# int s2/0
Router0(config - if)# ip add 16.16.1.1 255.255.255.0
Router0(config - if)# no shut
```

```
Router1(config)# int s2/0
Router1(config-if)# ip add 16.16.1.2 255.255.255.0
Router1(config-if)# clo ra 64000
Router1(config-if)# no shut
```

```
Router0# ping 16.16.1.2

Type escape sequence to abort.
Sending 5, 100-byte ICMP Echos to 16.16.1.2, timeout is 2 seconds:
!!!!!
Success rate is 100 percent (5/5), round-trip min/avg/max = 2/5/9 ms
```

(2) 查看路由器端口封装形式。

```
Router0# show interfaces s2/0
Serial2/0 is up, line protocol is up (connected)
  Hardware is HD64570
  Internet address is 16.16.1.1/24
  MTU 1500 bytes, BW 128 Kbit, DLY 20000 usec,
     reliability 255/255, txload 1/255, rxload 1/255
Encapsulation HDLC, loopback not set, keepalive set (10 sec)
… …
```

//默认封装形式为 HDLC。

```
Router1# show interfaces s2/0
Serial2/0 is up, line protocol is up (connected)
  Hardware is HD64570
  Internet address is 16.16.1.2/24
  MTU 1500 bytes, BW 128 Kbit, DLY 20000 usec,
     reliability 255/255, txload 1/255, rxload 1/255
Encapsulation HDLC, loopback not set, keepalive set (10 sec)
… …
```

(3) 改变端口 S2/0 封装形式。

```
Router0(config)# int s2/0
Router0(config-if)# encapsulation ?
  frame-relay  Frame Relay networks
  hdlc         Serial HDLC synchronous
  ppp          Point-to-Point protocol
```

//可以支持的封装形式。

```
Router0(config-if)# encapsulation ppp

% LINEPROTO-5-UPDOWN: Line protocol on Interface Serial2/0, changed state to down
Router0(config-if)# no shutdown
Router0(config-if)# end
Router0#
% SYS-5-CONFIG_I: Configured from console by console
```

```
Router0#ping 16.16.1.2

Type escape sequence to abort.
Sending 5, 100-byte ICMP Echos to 16.16.1.2, timeout is 2 seconds:
……
Success rate is 0 percent (0/5)
```

```
Router1(config)#int s2/0
Router1(config-if)#encapsulation ppp
Router1#show interfaces s2/0
Serial2/0 is up, line protocol is up (connected)
   Hardware is HD64570
   Internet address is 16.16.1.2/24
   MTU 1500 bytes, BW 128 Kbit, DLY 20000 usec,
      reliability 255/255, txload 1/255, rxload 1/255
Encapsulation PPP, loopback not set, keepalive set (10 sec)
… …
Router1#ping 16.16.1.1

Type escape sequence to abort.
Sending 5, 100-byte ICMP Echos to 16.16.1.1, timeout is 2 seconds:
!!!!!
Success rate is 100 percent (5/5), round-trip min/avg/max = 1/6/14 ms
```

//封装协议更改为PPP,链路通畅。

【提示】 改变其中一个串口的封装协议后,链路断开,说明两端封装协议需要匹配链路可以连通。

4.3 实验2 PAP认证实验

4.3.1 实验拓扑

路由器Router0和Router1串口连接,进行PAP认证,见图4.2。

设备名	IP地址	端口
Router0	16.16.1.1	S2/0
Router1	16.16.1.2	S2/0

4.3.2 实验内容

通过本实验可以掌握PAP认证原理和配置方法。
(1) 配置PPP封装形式。
(2) 配置PAP认证。

图4.2 路由器串口连接方式

4.3.3 实验步骤

首先配置路由器 Router0(远程路由器,被认证方)在路由器 Router1(中心路由器,认证方)取得验证。

(1) 两端路由器使用 PPP 封装形式。

```
Router0(config) # int s2/0
Router0(config - if) # encapsulation ppp
Router0(config - if) # ppp pap sent - username router0 password 123456
```

//在远程路由器 Router0 上,配置在中心路由器 Router1 登录的用户名和密码。

```
Router1(config) # int s2/0
Router1(config - if) # encapsulation ppp
Router1(config - if) # ppp authentication pap
```

//在 Router1 上配置 PAP 验证。

```
Router1(config) # username router0 password 123456
```

//以上步骤配置完成 Router0(远程路由器)在 Router1(中心路由器)取得验证,即单向验证。实际应用中可以使用双向验证,可以使用相似的配置完成。

【提示】 在 ISDN 拨号上网时,通常只是电信对用户进行验证(要根据用户名来收费),用户不能对电信进行验证,即验证是单向的。

(2) 查看认证过程。

```
Router0 # debug ppp authentication
PPP authentication debugging is on
```

//打开认证调试。

```
Router0(config - if) # shutdown
Router0(config - if) # no shut
```

//重新建立连接查看验证过程。

```
% LINK - 5 - CHANGED: Interface Serial2/0, changed state to up

Router0(config - if) #
Serial2/0 Using hostname from interface PAP

Serial2/0 Using password from interface PAP

Serial2/0 PAP: O AUTH - REQ id 17 len 15

Serial2/0 PAP: Phase is FORWARDING, Attempting Forward

% LINEPROTO - 5 - UPDOWN: Line protocol on Interface Serial2/0, changed state to up
```

//认证成功。

4.4 实验3 CHAP认证实验

4.4.1 实验拓扑

路由器Router0和Router1串口连接,进行CHAP认证,见图4.3。

设备名	IP地址	端口
Router0	16.16.1.1	S2/0
Router1	16.16.1.2	S2/0

4.4.2 实验内容

通过本实验可以掌握CHAP认证原理和配置方法。

图4.3 路由器串口连接方式

4.4.3 实验步骤

(1) 两端路由器使用PPP封装形式。

```
Router0(config)#int s2/0
Router0(config-if)#encapsulation ppp
```

```
Router1(config)#int s2/0
Router1(config-if)#encapsulation ppp
```

(2) 设置双方认证密码。

```
Router0(config)#username router1 password 123456
```

```
Router1(config)#username router0 password 123456
```

//注意,双方密码需要一致。

(3) 配置CHAP认证。

```
Router0(config)#int s2/0
Router0(config-if)#ppp authentication chap
```

```
Router1(config)#int s2/0
Router1(config-if)#ppp authentication chap
```

//CHAP配置过程中要求用户名为对方路由器名称,双方密码必须一致。原因是:由于CHAP默认使用本地路由器的名字作为建立PPP连接时的识别符。路由器在收到对方发送过来的询问消息后,将本地路由器的名字作为身份标识发送给对方;而在收到对方发过来的身份标识之后,默认使用本地验证方法,即在配置文件中寻找,看看有没有用户身份标识和密码;如果有,计算加密值,结果正确则验证通过;否则验证失败,连接无法建立。

【提示】 在配置验证时也可以选择同时使用 PAP 和 CHAP,如:

Router0(config - if)#**ppp authentication** *chap pap* 或
Router0(config - if)#**ppp authentication** *pap chap*

如果同时使用两种验证方式,那么在链路协商阶段将先用第一种验证方式进行验证。如果对方建议使用第二种验证方式或者只是简单拒绝使用第一种方式,那么将采用第二种方式。

4.5 实验 4 路由器配置为帧中继交换机

4.5.1 实验拓扑

路由器 Router0、Router1 和 Router2 串口连接,将 Router1 配置成帧中继交换机,见图 4.4。

设备名	IP 地址	端口	本地 DLCI
Router0	192.168.0.1	S2/0	102
Router2	172.16.0.1	S2/0	201

图 4.4 模拟帧中继交换机连接方式

4.5.2 实验内容

(1)理解帧中继交换机工作方式。
(2)使用路由器模拟帧中继交换机配置。

4.5.3 实验步骤

(1)首先配置帧中继交换机。

```
Router1 > enable
Router1 # conf t
Router1(config)# hostname FR
FR(config)# line console 0
FR(config - line)# exec - timeout 0 0
FR(config - line)# logging synchronous
```

//防止系统弹出信息打断命令的输入。

```
FR(config - line)# exit
FR(config)# no ip domain lookup
```

//防止输入错误命令进行 DNS 解析,影响工作效率。

```
FR(config)#end
```

(2) 配置 S2/0 口。

```
FR#conf t
FR(config)#frame-relay switching
FR(config)#interface s2/0
FR(config-if)#encapsulation frame-relay
FR(config-if)#clock rate 64000
FR(config-if)#frame-relay intf-type ?
  dce Configure a FR DCE
  dte Configure a FR DTE
  nni Configure a FR NNI
FR(config-if)#frame-relay intf-type dce
```

//设置帧中继的终端类型。

```
FR(config-if)#frame-relay lmi-type ?
  cisco
  ansi
  q933a
FR(config-if)#frame-relay lmi-type cisco
```

//配置帧中继的 LMI 接口类型，默认 cisco。

```
FR(config-if)#frame-relay route 102 interface serial3/0 201
```

//从 DLCI(102)进入 S2/0 的数据从 DLCI(201)(serial3/0)转发出去。

```
FR(config-if)#no shut
```

(3) 配置 S3/0 口。

```
FR(config-if)#interface serial3/0
FR(config-if)#encapsulation frame-relay
FR(config-if)#clock rate 64000
FR(config-if)#frame-relay lmi-type cisco
FR(config-if)#frame-relay intf-type dce
FR(config-if)#frame-relay route 201 interface serial2/0 102
FR(config-if)#no shut
FR(config-if)#end
```

//帧中继交换机配置完成。

【提示】 使用模拟器过程中，会有部分命令无法操作，可以更换另一种模拟器或者使用真实设备。

(4) 配置 Router0 和 Router2。

```
Router0(config)#interface s2/0
Router0(config-if)#ip add 192.168.0.1 255.255.255.0
Router0(config-if)#encapsulation frame-relay
Router0(config-if)#frame-relay lmi-type cisco
Router0(config-if)#frame-relay map ip 172.16.0.1 102 broadcast
```

//使用协议地址映射,即远端 IP 映射到本地的 DLCI,broadcast 参数指明路由更新信息通过该电路
//穿越网络(让 NBMA 支持广播)。
Router0(config-if)# no shutdown
Router0(config-if)# end

```
Router2(config)# interface serial2/0
Router2(config-if)# ip address 172.16.0.1 255.255.255.0
Router2(config-if)# encapsulation frame-relay
Router2(config-if)# frame-relay map ip 172.16.0.1 201 broadcast
Router2(config-if)# no shut
```

(5) 测试 Router0 和 Router2 之间的连通性。

```
Router0# ping 172.16.0.1

Type escape sequence to abort.
Sending 5, 100-byte ICMP Echos to 172.16.0.1, timeout is 2 seconds:
!!!!!
Success rate is 100 percent (5/5), round-trip min/avg/max = 1/6/14 ms
```

4.6 实验 5 帧中继点到点形式及帧中继映射

4.6.1 实验拓扑

路由器 Router0、Router1 和 Router2 通过帧中继交换机连接,实现 Router0 与 Router1、Router0 与 Router2 链路通信,见图 4.5。

设备名	IP 地址	端口	帧中继交换机端口	本地 DLCI
Router0	192.168.0.1	S2/0	S0	102/103
Router1	192.168.0.2	S2/0	S1	201
Router2	192.168.0.3	S2/0	S2	301

4.6.2 实验内容

(1) 帧中继的基本配置。
(2) 帧中继的动态映射。
(3) 帧中继的静态映射。

4.6.3 实验步骤

(1) 帧中继接口基本配置。

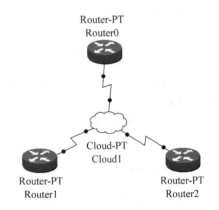

图 4.5 帧中继点到点及映射连接方式

```
Router0(config)# int s2/0
Router0(config-if)# ip add 192.168.0.1 255.255.255.0
Router0(config-if)# no shut
Router0(config-if)# encapsulation frame-relay
```

//使用命令"encapsulation frame-relay [*ietf*]"。帧中继有两种封装类型：*cisco* 和 *ietf*(Internet Engineering
//Task Force)。对于 Cisco 路由器，cisco 是它的默认值；对于非 Cisco 路由器，须选用 IETF 类型。

```
Router0(config-if)# frame-relay lmi-type cisco
```

//如果采用的是 Cisco 路由器且 IOS 是 11.2 及以后版本的，路由器可以自动适应 LMI 的类型，则本
//步骤可不做。国内帧中继线路一般采用 ansi 的 LMI 信令类型，这里采用的是 cisco。

```
Router1(config)# int s2/0
Router1(config-if)# ip add 192.168.0.2 255.255.255.0
Router1(config-if)# no shut
Router1(config-if)# encapsulation frame-relay
```

```
Router2(config)# int s2/0
Router2(config-if)# ip add 192.168.0.3 255.255.255.0
Router2(config-if)# no shut
Router2(config-if)# encapsulation frame-relay
```

（2）测试连通性。

```
Router0# ping 192.168.0.3

Type escape sequence to abort.
Sending 5, 100-byte ICMP Echos to 192.168.0.3, timeout is 2 seconds:
!!!!!
Success rate is 100 percent (5/5), round-trip min/avg/max = 6/7/10 ms

Router0# ping 192.168.0.2

Type escape sequence to abort.
Sending 5, 100-byte ICMP Echos to 192.168.0.2, timeout is 2 seconds:
!!!!!
Success rate is 100 percent (5/5), round-trip min/avg/max = 6/7/11 ms
```

//以上表示 Router0 与 Router1、Router0 与 Router2 链路通畅。

```
Router1# ping 192.168.0.3

Type escape sequence to abort.
Sending 5, 100-byte ICMP Echos to 192.168.0.3, timeout is 2 seconds:
.....
Success rate is 0 percent (0/5)
```

//由于 Router1 与 Router2 没有设置点对点通信，所以无法连通。

(3) 查看帧中继映射表。

```
Router0#show frame-relay map
Serial2/0 (up): ip 192.168.0.2 dlci 102, dynamic, broadcast, CISCO, status defined, active
Serial2/0 (up): ip 192.168.0.3 dlci 103, dynamic, broadcast, CISCO, status defined, active
```

//默认时,帧中继接口开启了动态映射,会自动建立帧中继映射,"dynamic"表明这是动态映射。

```
Router0#show frame-relay pvc

PVC Statistics for interface Serial2/0 (Frame Relay DTE)
DLCI = 102, DLCI USAGE = LOCAL, PVC STATUS = ACTIVE, INTERFACE = Serial2/0

  input pkts 14055         output pkts 32795        in bytes 1096228
  out bytes 6216155        dropped pkts 0           in FECN pkts 0
  in BECN pkts 0           out FECN pkts 0          out BECN pkts 0
  in DE pkts 0             out DE pkts 0
  out bcast pkts 32795     out bcast bytes 6216155

DLCI = 103, DLCI USAGE = LOCAL, PVC STATUS = ACTIVE, INTERFACE = Serial2/0

  input pkts 14055         output pkts 32795        in bytes 1096228
  out bytes 6216155        dropped pkts 0           in FECN pkts 0
  in BECN pkts 0           out FECN pkts 0          out BECN pkts 0
  in DE pkts 0             out DE pkts 0
  out bcast pkts 32795     out bcast bytes 6216155
```

(4) 手工配置帧中继映射。

```
Router0(config-if)#no frame-relay inverse-arp
```

//关闭自动映射。

```
Router0(config-if)#frame-relay map ip 192.168.0.2 102 broadcast
Router0(config-if)#frame-relay map ip 192.168.0.3 103 broadcast
```

```
Router1(config-if)#no frame-relay inverse-arp
Router1(config-if)#frame-relay map ip 192.168.0.1 201 broadcast
```

```
Router2(config-if)#no frame-relay inverse-arp
Router2(config-if)#frame-relay map ip 192.168.0.1 301 broadcast
```

```
Router0#show frame-relay lmi
LMI Statistics for interface Serial2/0 (Frame Relay DTE) LMI TYPE = CISCO
  Invalid Unnumbered info 0      Invalid Prot Disc 0
  Invalid dummy Call Ref 0       Invalid Msg Type 0
  Invalid Status Message 0       Invalid Lock Shift 0
  Invalid Information ID 0       Invalid Report IE Len 0
  Invalid Report Request 0       Invalid Keep IE Len 0
  Num Status Enq. Sent 379       Num Status msgs Rcvd 379
  Num Update Status Rcvd 0       Num Status Timeouts 16
```

从命令输出中可以得到的信息如下。

（1）LMI TYPE = CISCO：表明帧中继 LMI 类型为 cisco。

（2）Frame Relay DTE：这是帧中继 DTE。

（3）Num Status Enq. Sent 379：表明路由器向帧中继交换机发送的 LMI 状态查询消息的数量。

（4）Num Status msgs Rcvd 379：表明路由器从帧中继交换机收到的 LMI 状态信息的数量。

4.7 实验6 帧中继点到多点子接口

4.7.1 实验拓扑

路由器 Router0、Router1、Router2 和 Router3 通过帧中继交换机连接，实现 Router0 与 Router1、Router0 与 Router2、Router0 与 Router3 和 Router1 与 Router3 链路通信，Router0 端口设置复用，见图 4.6。

设备名	IP 地址	端口	帧中继交换机端口	本地 DLCI
Router0	192.168.0.1	S2/0.0	S0	102
Router0	192.168.0.1	S2/0.0	S0	103
Router0	172.16.0.1	S2/0.1	S0	104
Router1	192.168.0.2	S2/0	S1	201
Router2	192.168.0.3	S2/0	S2	301
Router3	172.16.1.1	S2/0	S3	401

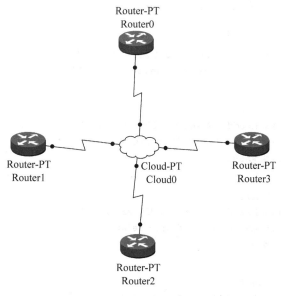

图 4.6 帧中继多点子接口连接方式

4.7.2 实验内容

掌握帧中继多点子接口控制方式,与路由协议相结合实现跨网络连通。
(1) 掌握点到多点子接口配置。
(2) 实现跨网络连通。

4.7.3 实验步骤

(1) 设置路由器帧中继封装协议。

```
Router0(config)#int s2/0
Router0(config-if)#no ip add
Router0(config-if)#encapsulation frame-relay
```

//在端口封装 frame-relay 协议。

```
Router0(config-if)#no shut
Router0(config-if)#exit
Router0(config)#int s2/0.1 multipoint
```

//设置子接口 s2/0.1 复用端口。

```
Router0(config-subif)#ip add 192.168.0.1 255.255.255.0
Router0(config-subif)#frame-relay map ip 192.168.0.2 102 broadcast
```

//设置静态映射。

```
Router0(config-subif)#frame-relay interface-dlci 103
```

//设置动态映射。

```
Router0(config-subif)#exit
Router0(config)#int s2/0.2 point-to-point
```

//设置子接口 s2/0.2 点对点端口。

```
Router0(config-subif)#ip add 172.16.0.1 255.255.255.0
Router0(config-subif)#frame-relay interface-dlci 104
```

```
Router1(config)#int s2/0
Router1(config-if)#ip add 192.168.0.2 255.255.255.0
Router1(config-if)#encapsulation frame-relay
Router1(config-if)#no shut
```

```
Router2(config)#int s2/0
Router2(config-if)#ip add 192.168.0.3 255.255.255.0
Router2(config-if)#encapsulation frame-relay
Router2(config-if)#no shut
```

```
Router3(config)#int s2/0
Router3(config-if)#ip add 172.16.0.2 255.255.255.0
Router3(config-if)#encapsulation frame-relay
Router3(config-if)#no shut
```

//Router1、Router2 和 Router3 作为远端协同路由器只需要设置帧中继封装协议。

【提示】 使用帧中继协议通常将一台路由器作为中心路由器,与远端路由器进行通信,不需要远端路由器之间通信,因此只需要对中心路由器配置。

(2) 数据通信验证。

```
Router1#ping 192.168.0.1

Type escape sequence to abort.
Sending 5, 100 - byte ICMP Echos to 192.168.0.1, timeout is 2 seconds:
!!!!!
Success rate is 100 percent (5/5), round - trip min/avg/max = 5/9/17 ms
```

```
Router3#ping 172.16.0.1

Type escape sequence to abort.
Sending 5, 100 - byte ICMP Echos to 172.16.0.1, timeout is 2 seconds:
!!!!!
Success rate is 100 percent (5/5), round - trip min/avg/max = 2/11/22 ms
```

//以上测试说明远端路由器与中心路由器连通。

```
Router3#ping 192.168.0.2

Type escape sequence to abort.
Sending 5, 100 - byte ICMP Echos to 192.168.0.2, timeout is 2 seconds:
.....
Success rate is 0 percent (0/5)
```

//路由器 Router1 与 Router3 无法连通,类似情况出现在 Router2 与它们之间。

(3) 结合路由协议完成配置。

```
Router0(config)#router rip
Router0(config - router)#net 192.168.0.0
Router0(config - router)#net 172.16.0.0
```

//使用简单的 RIP,后面章节详细阐述。

```
Router1(config)#router rip
Router1(config - router)#net 192.168.0.0
```

```
Router3(config)#router rip
Router3(config - router)#network 172.16.0.0
```

```
Router3#ping 192.168.0.2

Type escape sequence to abort.
Sending 5, 100 - byte ICMP Echos to 192.168.0.2, timeout is 2 seconds:
!!!!!
Success rate is 100 percent (5/5), round - trip min/avg/max = 8/15/18 ms
```

//再次测试,成功。

(4) 查看路由表和帧中继映射。

```
Router0 # show ip route
Codes: C - connected, S - static, I - IGRP, R - RIP, M - mobile, B - BGP
       D - EIGRP, EX - EIGRP external, O - OSPF, IA - OSPF inter area
       N1 - OSPF NSSA external type 1, N2 - OSPF NSSA external type 2
       E1 - OSPF external type 1, E2 - OSPF external type 2, E - EGP
       i - IS-IS, L1 - IS-IS level-1, L2 - IS-IS level-2, ia - IS-IS inter area
       * - candidate default, U - per-user static route, o - ODR
       P - periodic downloaded static route

Gateway of last resort is not set

     172.16.0.0/24 is subnetted, 1 subnets
C       172.16.0.0 is directly connected, Serial2/0.2
C    192.168.0.0/24 is directly connected, Serial2/0.1

Router0 # show frame-relay map
Serial2/0.1 (up): ip 192.168.0.2 dlci 102, static, broadcast, CISCO, status defined, active
Serial2/0.1 (up): ip 192.168.0.3 dlci 103, dynamic, broadcast, CISCO, status defined, active
Serial2/0.2 (up): point-to-point dlci, dlci 104, broadcast, status defined, active
```

第 5 章　网络安全配置

5.1　基础知识

5.1.1　ACL 概述

路由器提供了基本的流量过滤能力,最简单方便且易于理解和使用的,就是访问控制列表(Access Control List,ACL)。ACL 就是用来在使用路由技术的网络里,识别和过滤那些由某些网络发出的或者被发送出去某些网络的符合所规定条件的数据流量,以决定这些数据流量是应该转发还是应该丢弃的技术。

ACL 是一个连续的允许和拒绝语句的集合,关系到地址或上层协议。在使用 ACL 时,是把预先定义好的 ACL 放置在路由器的接口上,对接口上进方向或者出方向的数据包进行过滤。但是 ACL 只能过滤经过路由器的数据包,对于路由器自己本身所产生的数据包,应用在接口上的 ACL 是不能过滤的。

ACL 通过在路由器接口处控制被路由的分组是被转发还是被阻塞来过滤网络流量。路由器基于 ACL 中指定的条件来决定转发还是丢弃分组。ACL 中的条件可以是流量的源地址、流量的目的地址、上层协议、端口号或应用等。

1. 标准 ACL

标准 ACL 检查可以被路由的 IP 分组的源地址并且把它与 ACL 中的条件判断语句相比较。标准 ACL 可以根据网络、子网或主机 IP 地址允许或拒绝整个协议组(如 IP)。

2. 扩展 ACL

扩展 ACL 比标准 ACL 使用得更多,因为它提供了更大的弹性和控制范围。扩展 ACL 既可检查分组的源地址和目的地址,也检查协议类型和 TCP 或 UDP 的端口号。

扩展 ACL 可以基于分组的源地址、目的地址、协议类型、端口地址和应用来决定访问是被允许或者被拒绝。扩展 ACL 比标准 ACL 提供了更广阔的控制范围和更多的分组处理方法。

3. 命名 ACL

以列表名称代替列表编号来定义 ACL,同样包括标准和扩展两种列表。

5.1.2　ACL 在网络中的应用位置

在一个网络中,ACL 是通过过滤分组和拒绝那些不需要的通信流量来实现控制的。放置 ACL 需要考虑的一个重要条件是应该在什么地方放置 ACL。当 ACL 被放置在正确的地方时,它不仅可以过滤通信量,而且可以使整个网络更有效地运作。为了考虑通信流量,

ACL 应该放置在对网络增长影响最大的地方。

放置 ACL 的一般原则是，尽可能把扩展 ACL 放置在距离要被拒绝的通信量近的地方。标准 ACL 由于不能指定目的地址，所以它们应该尽可能放置在距离目的地最近的地方。

在访问控制列表的学习中，要特别注意以下两个术语。

(1) 通配符掩码：一个 32b 长的数字字符串，它规定了当一个 IP 地址与其他的 IP 地址进行比较时，该 IP 地址中哪些位应该被忽略。通配符掩码中的"1"表示忽略 IP 地址中对应的位，而"0"则表示该位必须匹配。两种特殊的通配符掩码是"255.255.255.255"和"0.0.0.0"，前者等价于关键字"any"，而后者等价于关键字"host"。

(2) Inbound 和 Outbound：当在接口上应用访问控制列表时，用户要指明访问控制列表是应用于流入数据还是流出数据。

总之，ACL 的应用非常广泛，它可以实现如下的功能：

(1) 拒绝或允许流入(或流出)的数据流通过特定的接口。

(2) 为 DDR 应用定义感兴趣的数据流。

(3) 过滤路由更新的内容。

(4) 控制对虚拟终端的访问。

(5) 提供流量控制。

5.1.3　NAT 概述

网络地址转换(Network Address Translation，NAT)属接入广域网(WAN)技术，是一种将私有(保留)地址转化为合法 IP 地址的转换技术，它被广泛应用于各种类型 Internet 接入方式和各种类型的网络中。原因很简单，NAT 不仅完美地解决了 IP 地址不足的问题，而且还能够有效地避免来自网络外部的攻击，隐藏并保护网络内部的计算机。

NAT 的实现方式有三种，即静态转换 Static Nat、动态转换 Dynamic Nat 和端口多路复用 OverLoad。

(1) 静态转换是指将内部网络的私有 IP 地址转换为公有 IP 地址，IP 地址对是一对一的，是一成不变的，某个私有 IP 地址只转换为某个公有 IP 地址。借助于静态转换，可以实现外部网络对内部网络中某些特定设备(如服务器)的访问。

(2) 动态转换是指将内部网络的私有 IP 地址转换为公用 IP 地址时，IP 地址对是不确定的，而是随机的，所有被授权访问上 Internet 的私有 IP 地址可随机转换为任何指定的合法 IP 地址。也就是说，只要指定哪些内部地址可以进行转换，以及用哪些合法地址作为外部地址时，就可以进行动态转换。动态转换可以使用多个合法外部地址集。当 ISP 提供的合法 IP 地址略少于网络内部的计算机数量时。可以采用动态转换的方式。

(3) 端口多路复用是指改变外出数据包的源端口并进行端口转换，即端口地址转换(Port Address Translation，PAT)。采用端口多路复用方式，内部网络的所有主机均可共享一个合法的外部 IP 地址实现对 Internet 的访问，从而可以最大限度地节约 IP 地址资源。同时，又可隐藏网络内部的所有主机，有效避免来自 Internet 的攻击。因此，目前网络中应用最多的就是端口多路复用方式。

5.2 实验1 标准ACL实验

5.2.1 实验拓扑

通过对路由器Router0的配置，使得主机PC1不能访问位于192.168.0.0网段主机PC0，见图5.1。

设备名	IP地址	端口
Router0	192.168.0.1	F0/0
Router0	172.16.0.1	F1/0
PC0	192.168.0.2	F0
PC1	172.16.0.2	F0
PC2	172.16.0.3	F0

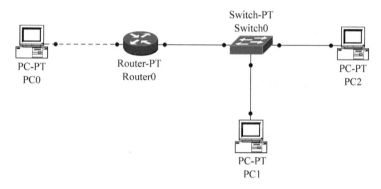

图5.1 标准ACL连接方式

5.2.2 实验内容

通过本实验可以掌握标准访问控制列表的原理和配置方法。
（1）配置标准访问控制列表。
（2）测试配置结果。

5.2.3 实验步骤

（1）配置路由器端口地址。

```
Router0(config)#int f0/0
Router0(config-if)#ip add 192.168.0.1 255.255.255.0
Router0(config-if)#no shut
Router0(config-if)#int f1/0
Router0(config-if)#ip add 172.16.0.1 255.255.255.0
Router0(config-if)#no shut
```

(2) 测试目前连通情况。

```
PC0 > ping 172.16.0.2

Pinging 172.16.0.2 with 32 bytes of data:

Reply from 172.16.0.2: bytes = 32 time = 5ms TTL = 127
Reply from 172.16.0.2: bytes = 32 time = 5ms TTL = 127
Reply from 172.16.0.2: bytes = 32 time = 13ms TTL = 127
Reply from 172.16.0.2: bytes = 32 time = 6ms TTL = 127

Ping statistics for 172.16.0.2:
    Packets: Sent = 4, Received = 4, Lost = 0 (0% loss),
Approximate round trip times in milli-seconds:
Minimum = 5ms, Maximum = 13ms, Average = 7ms
```

//连通成功。

(3) 配置标准的访问列表。

```
Router0(config)#access-list 1 deny 172.16.0.2 0.0.0.0
Router0(config)#access-list 1 permit any
```

//为控制列表"1"添加访问规则。

(4) 在接口上应用该访问列表。

```
Router0(config)#int f0/0
Router0(config-if)#ip access-group 1 out
```

//为端口分配控制列表"1",并指定数据流方向。

(5) 测试配置结果。

```
PC0 > ping 172.16.0.2

Pinging 172.16.0.2 with 32 bytes of data:

Request timed out.
Request timed out.
Request timed out.
Request timed out.

Ping statistics for 172.16.0.2:
Packets: Sent = 4, Received = 0, Lost = 4 (100% loss)
```

//因为控制列表的存在,无法连通 PC1,但 PC2 仍可以连通。

```
Router0#show ip access-lists
Standard IP access list 1
    deny host 172.16.0.2 (4 match(es))
    permit any (4 match(es))
```

//以上输出表明路由器 Router0 上定义的标准访问控制列表为"1",括号中的数字表示匹配条件的数
//据包的个数,可以使用"clear access-list counters"将访问控制列表计数器清零。

【提示】

(1) ACL 定义好，可以在很多地方应用，接口上应用只是其中之一，其他的常用应用包括在 route map 中的 match 应用和在 vty 下用"access-class"命令调用，来控制 Telnet 的访问；

(2) 访问控制列表表项的检查按自上而下的顺序进行，并且从第一个表项开始，所以必须考虑在访问控制列表中定义语句的次序；

(3) 路由器不对自身产生的 IP 数据包进行过滤；

(4) 访问控制列表最后一条是隐含的拒绝所有；

(5) 每一个路由器接口的每一个方向，每一种协议只能创建一个 ACL；

(6) "access-class"命令只对标准 ACL 有效。

5.3 实验 2　扩展访问控制列表

5.3.1 实验拓扑

通过对路由器 Router0、Router1 和 Router2 的配置，要求只允许 PC1 所在网段的主机访问路由器 Router1 的 WWW 和 Telnet 服务，并拒绝 PC2 所在网段 ping 路由器 Router1，见图 5.2。

设备名	IP 地址	端口
Router0	192.168.0.1	F0/0
Router0	192.168.1.1	F1/0
Router0	16.16.1.1	S2/0
Router1	16.16.1.2	S2/0
Router1	16.16.2.1	S3/0
Router1	192.168.3.1	F0/0
Router2	16.16.2.2	S2/0
Router2	192.168.2.1	F0/0
PC0	192.168.0.2	F0
PC1	192.168.1.2	F0
PC2	192.168.2.2	F0
Server0	192.168.3.2	F0

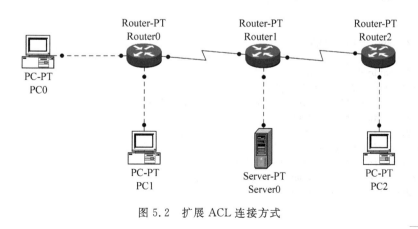

图 5.2　扩展 ACL 连接方式

5.3.2 实验内容

(1) 定义扩展访问控制列表。
(2) 应用扩展访问控制列表。
(3) 扩展访问控制列表调试。

5.3.3 实验步骤

(1) 使用 RIP 路由协议连通整个网络。

```
Router0(config)# int s2/0
Router0(config-if)# ip add 16.16.1.1 255.255.255.0
Router0(config-if)# no shut
Router0(config-if)# int f0/0
Router0(config-if)# ip add 192.168.0.1 255.255.255.0
Router0(config-if)# no shut
Router0(config-if)# int f1/0
Router0(config-if)# ip add 192.168.1.1 255.255.255.0
Router0(config-if)# no shut
Router0(config-if)# exit
Router0(config)# router rip
Router0(config-router)# net 192.168.0.0
Router0(config-router)# net 192.168.1.0
Router0(config-router)# net 16.16.1.0
Router0(config-router)# exit
```

```
Router1(config)# int s2/0
Router1(config-if)# ip add 16.16.1.2 255.255.255.0
Router1(config-if)# clock rate 64000
Router1(config-if)# no shut
Router1(config-if)# int s3/0
Router1(config-if)# ip add 16.16.2.1 255.255.255.0
Router1(config-if)# clock rate 64000
Router1(config-if)# no shut
Router1(config-if)# exit
Router1(config)# router rip
Router1(config-router)# net 16.16.0.0
```

```
Router2(config)# int s2/0
Router2(config-if)# ip add 16.16.2.2 255.255.255.0
Router2(config-if)# no shut
Router2(config-if)# int f0/0
Router2(config-if)# ip add 192.168.2.1 255.255.255.0
Router2(config-if)# no shut
Router2(config-if)# exit
Router2(config)# router rip
Router2(config-router)# net 16.16.0.0
Router2(config-router)# net 192.168.2.0
Router2(config-router)# end
```

```
PC0 > ping 192.168.2.2

Pinging 192.168.2.2 with 32 bytes of data:

Reply from 192.168.2.2: bytes = 32 time = 14ms TTL = 125
Reply from 192.168.2.2: bytes = 32 time = 16ms TTL = 125
Reply from 192.168.2.2: bytes = 32 time = 7ms TTL = 125
Reply from 192.168.2.2: bytes = 32 time = 15ms TTL = 125

Ping statistics for 192.168.2.2:
    Packets: Sent = 4, Received = 4, Lost = 0 (0% loss),
Approximate round trip times in milli - seconds:
Minimum = 7ms, Maximum = 16ms, Average = 13ms
```

//经过测试整个网络连通。

(2) 配置路由器 Router0。

```
Router0(config) # access - list 100 permit tcp 192.168.1.0 0.0.0.255 host 192.168.3.2 eq www
Router0(config) # access - list 100 permit tcp 192.168.1.0 0.0.0.255 host 16.16.1.2 eq www
Router0(config) # access - list 100 permit tcp 192.168.1.0 0.0.0.255 host 16.16.2.1 eq www
Router0(config) # access - list 100 permit tcp 192.168.1.0 0.0.0.255 host 192.168.3.2 eq telnet
Router0(config) # access - list 100 permit tcp 192.168.1.0 0.0.0.255 host 16.16.1.2 eq telnet
Router0(config) # access - list 100 permit tcp 192.168.1.0 0.0.0.255 host 16.16.2.1 eq telnet
Router0(config) # int f0/0
Router0(config - if) # ip access - group 100 in
```

(3) 配置路由器 Router1。

```
Router1(config) # ip name - server 192.168.3.2
```

//配置域名服务器主机。

```
Router1(config) # line vty 0 4
Router1(config - line) # password cisco
Router1(config - line) # login
```

(4) 配置路由器 Router2。

```
Router2(config) # access - list 101 deny icmp 192.168.2.0 0.0.0.255 host 192.168.3.2
Router2(config) # access - list 101 deny icmp 192.168.2.0 0.0.0.255 host 16.16.1.2
Router2(config) # access - list 101 deny icmp 192.168.2.0 0.0.0.255 host 16.16.2.1
Router2(config) # access - list 101 permit ip any any
Router2(config) # int fa0/0
Router2(config - if) # ip acc
Router2(config - if) # ip access - group 101 in
```

【提示】 尽量考虑将扩展的访问控制列表放在靠近过滤源的位置上,这样创建的过滤器就不会反过来影响其他接口上的数据流。另外,尽量使标准的访问控制列表靠近目的,由于标准访问控制列表只使用源地址,如果将其靠近源会阻止数据包流向其他端口。

(5) 验证调试。

```
PC1 > telnet 192.168.3.2
Trying 192.168.3.2 ...
% Connection refused by remote host
PC > telnet 16.16.1.2
Trying 16.16.1.2 ...Open


User Access Verification

Password:
router1 > exit
```

//PC1 所在网络远程访问成功。

```
PC2 > ping 192.168.3.2

Pinging 192.168.3.2 with 32 bytes of data:

Request timed out.
Request timed out.
Request timed out.
Request timed out.

Ping statistics for 192.168.3.2:
Packets: Sent = 4, Received = 0, Lost = 4 (100% loss)
```

//PC2 所在网络无法连通 Router1 服务器。

```
Router2#show ip access - lists
Extended IP access list 101
    deny icmp 192.168.2.0 0.0.0.255 host 192.168.3.2 (4 match(es))
    deny icmp 192.168.2.0 0.0.0.255 host 16.16.1.2
    deny icmp 192.168.2.0 0.0.0.255 host 16.16.2.1 (4 match(es))
    permit ip any any
```

//以上输出表明扩展访问控制列表"101"拒绝来自网络"192.168.2.0"的数据信息。

5.4 实验3 命名访问控制列表

5.4.1 实验拓扑

要求同上,命名 ACL 连接方式,见图 5.3。

设备名	IP 地址	端口
Router0	192.168.0.1	F0/0
Router0	192.168.1.1	F1/0
Router0	16.16.1.1	S2/0
Router1	16.16.1.2	S2/0
Router1	16.16.2.1	S3/0
Router1	192.168.3.1	F0/0
Router2	16.16.2.2	S2/0
Router2	192.168.2.1	F0/0
PC0	192.168.0.2	F0
PC1	192.168.1.2	F0
PC2	192.168.2.2	F0
Server0	192.168.3.2	F0

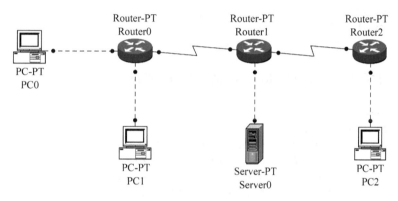

图 5.3 命名 ACL 连接方式

5.4.2 实验内容

命名 ACL 允许在标准 ACL 和扩展 ACL 中,使用字符串代替前面所使用的数字来表示 ACL。命名 ACL 还可以被用来从某一特定的 ACL 中删除个别的控制条目,这样可以让网络管理员方便地修改 ACL。

(1) 定义命名访问控制列表。
(2) 应用命名访问控制列表。

5.4.3 实验步骤

(1) 在路由器 Router0 和 Router2 上配置命名的扩展 ACL。

```
Router0(config)# ip access-list extended a1
```

//定义命名访问控制列表 a1。

```
Router0(config-ext-nacl)#permit tcp 192.168.1.0 0.0.0.255 host 16.16.1.2 eq www
Router0(config-ext-nacl)#permit tcp 192.168.1.0 0.0.0.255 host 192.168.3.2 eq www
Router0(config-ext-nacl)#permit tcp 192.168.1.0 0.0.0.255 host 16.16.2.1 eq www
Router0(config-ext-nacl)#permit tcp 192.168.1.0 0.0.0.255 host 192.168.3.2 eq telnet
Router0(config-ext-nacl)#permit tcp 192.168.1.0 0.0.0.255 host 16.16.1.2 eq telnet
Router0(config-ext-nacl)#permit tcp 192.168.1.0 0.0.0.255 host 16.16.2.1 eq telnet
Router0(config-ext-nacl)#exit
Router0(config)#int f0/0
Router0(config-if)#ip access-group a1 in
```

```
Router2(config)#ip access-list extended a2
```

//定义命名访问控制列表 a2。

```
Router2(config-ext-nacl)#deny icmp 192.168.2.0 0.0.0.255 host 192.168.3.2
Router2(config-ext-nacl)#deny icmp 192.168.2.0 0.0.0.255 host 16.16.1.2
Router2(config-ext-nacl)#deny icmp 192.168.2.0 0.0.0.255 host 16.16.2.1
Router2(config-ext-nacl)#permit ip any any
Router2(config-ext-nacl)#exit
Router2(config)#int f0/0
Router2(config-if)#ip access-group a2 in
```

(2) 查看路由器 Router0 和 Router2 的命名访问控制列表。

```
Router0#show access-lists
Extended IP access list a1
    permit tcp 192.168.1.0 0.0.0.255 host 16.16.1.2 eq www
    permit tcp 192.168.1.0 0.0.0.255 host 192.168.3.2 eq www
    permit tcp 192.168.1.0 0.0.0.255 host 16.16.2.1 eq www
    permit tcp 192.168.1.0 0.0.0.255 host 192.168.3.2 eq telnet
    permit tcp 192.168.1.0 0.0.0.255 host 16.16.1.2 eq telnet
permit tcp 192.168.1.0 0.0.0.255 host 16.16.2.1 eq telnet
```

```
Router2#show access-lists
Extended IP access list a2
    deny icmp 192.168.2.0 0.0.0.255 host 192.168.3.2
    deny icmp 192.168.2.0 0.0.0.255 host 16.16.1.2
    deny icmp 192.168.2.0 0.0.0.255 host 16.16.2.1
permit ip any any
```

5.5 实验 4 访问控制列表综合应用

5.5.1 实验拓扑

路由器配置动态路由协议,使整个网络连通,然后建立访问控制列表,实现应用要求,见图 5.4。

设备名	IP 地址	端口
Router0	192.168.0.1	F0/0
Router0	192.168.1.1	F1/0
Router0	16.16.1.1	S2/0
Router1	16.16.1.2	S2/0
Router1	192.168.2.1	F0/0
PC0	192.168.0.2	F0
PC1	192.168.1.2	F0
PC2	192.168.2.2	F0

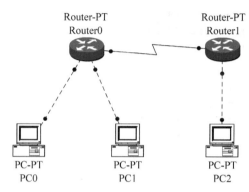

图 5.4　ACL 综合应用

5.5.2　实验内容

（1）配置动态路由协议。

（2）建立访问控制列表。

（3）验证调试。

5.5.3　实验步骤

（1）配置设备 IP 地址。

```
Router0(config)# int f0/0
Router0(config-if)# ip add 192.168.0.1 255.255.255.0
Router0(config-if)# no shut
Router0(config-if)# int f1/0
Router0(config-if)# ip add 192.168.1.1 255.255.255.0
Router0(config-if)# no shut
Router0(config-if)# int s2/0
Router0(config-if)# ip add 16.16.1.1 255.255.255.0
Router0(config-if)# no shut
```

```
Router1(config)# int s2/0
Router1(config-if)# ip add 16.16.1.2 255.255.255.0
Router1(config-if)# clock rate 64000
```

```
Router1(config-if)#no shut
Router1(config-if)#int f0/0
Router1(config-if)#ip add 192.168.2.1 255.255.255.0
Router1(config-if)#no shut
```

(2) 配置动态路由协议。

```
Router0(config)#router rip
Router0(config-router)#version 2
```

//选择协议版本,后面章节详细阐述。

```
Router0(config-router)#net 192.168.0.0
Router0(config-router)#net 192.168.1.0
Router0(config-router)#net 16.16.1.0
```

```
Router1(config)#router rip
Router1(config-router)#v 2
Router1(config-router)#net 192.168.2.0
Router1(config-router)#net 16.16.1.0
```

```
PC0>ping 192.168.2.2

Pinging 192.168.2.2 with 32 bytes of data:

Reply from 192.168.2.2: bytes=32 time=12ms TTL=126
Reply from 192.168.2.2: bytes=32 time=6ms TTL=126
Reply from 192.168.2.2: bytes=32 time=16ms TTL=126
Reply from 192.168.2.2: bytes=32 time=14ms TTL=126

Ping statistics for 192.168.2.2:
    Packets: Sent = 4, Received = 4, Lost = 0 (0% loss),
Approximate round trip times in milli-seconds:
Minimum = 6ms, Maximum = 16ms, Average = 12ms
```

//连通成功。

(3) 配置标准访问列表:允许源网络地址为 192.168.0.0 的数据通过,但拒绝其中的主机 PC1。

```
Router0(config)#access-list 1 deny host 192.168.1.2
Router0(config)#access-list 1 permit 192.168.0.0 0.0.255.255
Router0(config)#int s2/0
Router0(config-if)#ip access-group 1 out
```

```
PC1>ping 192.168.2.1

Pinging 192.168.2.1 with 32 bytes of data:
```

```
Reply from 192.168.1.1: Destination host unreachable.
Reply from 192.168.1.1: Destination host unreachable.
Reply from 192.168.1.1: Destination host unreachable.
Reply from 192.168.1.1: Destination host unreachable.

Ping statistics for 192.168.2.1:
Packets: Sent = 4, Received = 0, Lost = 4 (100% loss)
```

//测试成功。

(4) 配置扩展访问控制列表,禁止主机 PC0 远程登录路由器 Router1。

```
Router0(config)# int s2/0
Router0(config-if)# no ip access-group 1 out
```

//从接口上删除 ACL。

```
Router0(config)# access-list 110 deny tcp host 192.168.0.2 host 192.168.2.1 eq telnet
Router0(config)# access-list 110 permit ip any any
Router0(config)# int s2/0
Router0(config-if)# ip access-group 110 out
```

```
Router1(config)# line vty 0 4
Router1(config-line)# password cisco
Router1(config-line)# login
```

//路由器 Router1 开启远程登录。

```
PC0> telnet 192.168.2.1
Trying 192.168.2.1 ...
% Connection timed out; remote host not responding
```

//测试成功。

(5) 配置命名访问控制列表,只允许主机 PC0 远程登录路由器 Router1。

```
Router0(config)# ip access-list extended a1
Router0(config-ext-nacl)# permit tcp host 192.168.0.2 host 192.168.2.1 eq telnet
Router0(config-ext-nacl)# exit
Router0(config)# int s2/0
Router0(config-if)# ip access-group a1 out
```

```
PC0> telnet 192.168.2.1
Trying 192.168.2.1 ...Open


User Access Verification

Password:
router1> exit
```

```
[Connection to 192.168.2.1 closed by foreign host]
PC1 > telnet 192.168.2.1
Trying 192.168.2.1 ...
% Connection timed out; remote host not responding
```
//测试成功。

5.6 实验5 静态 NAT 服务

5.6.1 实验拓扑

配置路由器 Router0 提供地址转换，实现网络连通，见图 5.5。

设备名	IP 地址	端口
Router0	192.168.0.1	F0/0
Router0	16.16.1.1	S2/0
Router1	16.16.1.2	S2/0
Router1	192.168.1.1	F0/0
PC0	192.168.0.2	F0
PC1	192.168.0.3	F0
Server0	192.168.1.2	F0

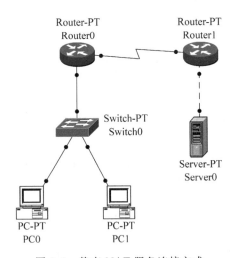

图 5.5 静态 NAT 服务连接方式

5.6.2 实验内容

掌握静态 NAT 的特征和配置方法。
(1) 配置 NAT 服务。
(2) 验证调试。

5.6.3 实验步骤

(1) 配置路由器 Router0 的 NAT 服务。

```
Router0(config)#router rip
Router0(config-router)#net 16.16.1.0
Router0(config-router)#exit
Router0(config)#ip nat inside source static 192.168.0.2 16.16.1.3
```

//配置静态 NAT 映射。

```
Router0(config)#ip nat inside source static 192.168.0.3 16.16.1.4
Router0(config)#int f0/0
Router0(config-if)#ip nat inside
```

//配置 NAT 内部接口。

```
Router0(config-if)#int s2/0
Router0(config-if)#ip nat out
```

//配置 NAT 外部接口。

```
Router0(config-if)#ip nat outside
```

(2) 配置路由器 Router1。

```
Router1(config)#router rip
Router1(config-router)#net 16.16.1.0
Router1(config-router)#net 192.168.1.0
```

【提示】 在路由器 Router0 上没有配置网络 192.168.0.0 的路由协议,使用 NAT 服务实现网络连通。

(3) 验证调试。

```
Router0#debug ip nat
IP NAT debugging is on
Router0#
NAT: s = 192.168.0.2 -> 16.16.1.3, d = 192.168.1.2 [13]

NAT*: s = 192.168.1.2, d = 16.16.1.3 -> 192.168.0.2 [21]

NAT: s = 192.168.0.2 -> 16.16.1.3, d = 192.168.1.2 [14]

NAT*: s = 192.168.1.2, d = 16.16.1.3 -> 192.168.0.2 [22]

NAT: s = 192.168.0.2 -> 16.16.1.3, d = 192.168.1.2 [15]

NAT*: s = 192.168.1.2, d = 16.16.1.3 -> 192.168.0.2 [23]

NAT: s = 192.168.0.2 -> 16.16.1.3, d = 192.168.1.2 [16]

NAT*: s = 192.168.1.2, d = 16.16.1.3 -> 192.168.0.2 [24]
```

//以上输出表明了 NAT 的转换过程。首先把私有地址"192.168.0.2"转换成公网地址"16.16.1.3"
//访问地址"192.168.1.2",然后回来的时候把公网地址"16.16.1.3"转换成私有地址"192.168.0.2"。

【提示】 开启调试功能后需要主机 PC0 向服务器 Server0 发送数据才可以回显调试信息。

(4) 查看 NAT 转换。

```
Router0#show ip nat translations
Pro   Inside global    Inside local    Outside local    Outside global
---   16.16.1.3        192.168.0.2     ---              ---
---   16.16.1.4        192.168.0.3     ---              ---
```

//输出表明了内部全局地址和内部局部地址的对应关系。

【提示】

① 内部局部(inside local)地址:在内部网络使用的地址,往往是主机网卡地址。

② 内部全局(inside global)地址:用来代替一个或多个本地 IP 地址的、对外的、向 NIC 注册过的地址。

③ 外部局部(outside local)地址:一个外部主机相对于内部网络所用的 IP 地址。不一定是合法的地址。

④ 外部全局(outside global)地址:外部网络主机的合法 IP 地址。

5.7 实验 6 动态 NAT 与 PAT 服务

5.7.1 实验拓扑

配置路由器 Router0 提供多种 NAT 服务方式,实现网络连通,见图 5.6。

设备名	IP 地址	端口
Router0	192.168.0.1	F0/0
Router0	16.16.1.1	S2/0
Router1	16.16.1.2	S2/0
Router1	192.168.1.1	F0/0
Router1	16.16.2.1	S3/0
Router2	192.168.2.1	F0/0
Router2	16.16.2.2	S2/0
PC0	192.168.0.2	F0
PC1	192.168.0.3	F0
PC2	192.168.2.2	F0
PC3	192.168.2.3	F0
Server0	192.168.1.2	F0

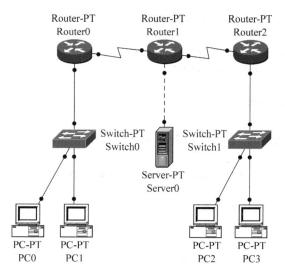

图 5.6 动态 NAT 服务连接方式

5.7.2 实验内容

掌握动态 NAT 和 PAT 配置方法。

(1) 配置动态 NAT 服务。

(2) 掌握 overload 的使用。

(3) 配置 PAT 服务。

5.7.3 实验步骤

(1) 配置路由器端口的 IP 地址,并在 Router0 和 Router2 上配置到 192.168.1.0 网段动态路由协议。

```
Router0(config)#router rip
Router0(config-router)#net 16.16.1.0
```

```
Router2(config)#router rip
Router2(config-router)#net 16.16.2.0
```

(2) 在 Router1 上分别为 Router0 和 Router2 配置一条到达内部全局地址的动态路由协议。

```
Router1(config)#int s2/0
Router1(config-if)#ip add 16.16.1.2 255.255.255.0
Router1(config-if)#clock rate 64000
Router1(config-if)#no shut
Router1(config-if)#exit
Router1(config)#router rip
Router1(config-router)#net 16.16.1.0
Router1(config-router)#net 16.16.2.0
```

```
Router1(config-router)#net 192.168.1.0
Router1(config)#int s3/0
Router1(config-if)#ip add 16.16.2.1 255.255.255.0
Router1(config-if)#clock rate 64000
Router1(config-if)#no shut
Router1(config-if)#exit
```

(3) 在 Router0 的 f0/0 端口配置 ACL 过滤掉所有私有地址。

```
Router0(config)#ip nat pool cisco1 16.16.1.1 16.16.1.254 net 255.255.255.0
```

//定义动态 NAT 转换地址池 IP 地址范围。

```
Router0(config)#access-list 2 permit 192.168.0.0 0.0.0.255
Router0(config)#ip nat ?
  inside    Inside address translation
  outside   Outside address translation
  pool      Define pool of addresses
```

//可选择转换方式。

```
Router0(config)#ip nat inside source list 2 pool cisco1
```

//配置动态 NAT 映射。

```
Router0(config)#int fa0/0
Router0(config-if)#ip nat inside
Router0(config-if)#int s2/0
Router0(config-if)#ip nat outside
Router0(config-if)#end
```

(4) 测试动态 NAT 配置。

```
Router0#debug ip nat
IP NAT debugging is on
Router0#
NAT: s=192.168.0.2->16.16.1.1, d=192.168.1.2 [13]

NAT*: s=192.168.1.2, d=16.16.1.1->192.168.0.2 [34]

NAT: s=192.168.0.2->16.16.1.1, d=192.168.1.2 [14]

NAT*: s=192.168.1.2, d=16.16.1.1->192.168.0.2 [35]

NAT: s=192.168.0.2->16.16.1.1, d=192.168.1.2 [15]

NAT*: s=192.168.1.2, d=16.16.1.1->192.168.0.2 [36]
```

【提示】 如果动态 NAT 地址池中没有足够的地址作动态映射,则会出现类似下面的信息,提示 NAT 转换失败,并丢弃数据包。

```
NAT: translation failed (A), dropping packet s = 192.168.0.2 d = 192.168.1.2
Router0#show ip nat statistics
Total translations: 7 (0 static, 7 dynamic, 7 extended)
```

//共有 7 个动态转换。

```
Outside Interfaces: Serial2/0
```

//NAT 外部接口。

```
Inside Interfaces: FastEthernet0/0
Hits: 19 Misses: 55
Expired translations: 12
Dynamic mappings:
```

//动态映射。

```
-- Inside Source
access-list 2 pool cisco1 refCount 7
pool cisco1: netmask 255.255.255.0
```

//地址池名字和掩码。

```
start 16.16.1.1 end 16.16.1.254
```

//地址池范围。

```
type generic, total addresses 254, allocated 1 (0%), misses 0
```

//共 254 个地址,分配 1 个。

```
Router0#show ip nat translations
Pro    Inside global      Inside local     Outside local    Outside global
icmp   16.16.1.1:25       192.168.0.2:25   192.168.1.2:25   192.168.1.2:25
icmp   16.16.1.1:26       192.168.0.2:26   192.168.1.2:26   192.168.1.2:26
```

//动态 NAT 转换信息。

【提示】 查看 NAT 转换信息与统计信息需要在该路由器转换工作状态下进行,即需要主机进行相应访问操作。

(5) 配置 PAT 服务。

```
Router0(config)#ip nat pool cisco1 16.16.1.1 16.16.1.254 net 255.255.255.0
Router0(config)#access-list 2 permit 192.168.0.0 0.0.0.255
Router0(config)#ip nat inside source list 2 pool cisco1 overload
ipnat_add_dynamic_cfg: id 2, flag 5, range 0

poolstart 16.16.1.1 poolend 16.16.1.254

Router0(config)# id 2, flags 0, domain 0, lookup 0, aclnum 2 ,
       aclname 2 , mapname idb 0
```

//配置 PAT 服务的返回信息。

```
Router0(config)# int fa0/0
Router0(config-if)# ip nat inside
Router0(config-if)# int s2/0
Router0(config-if)# ip nat outside
Router0(config-if)# end
Router0# debug ip nat
IP NAT debugging is on
Router0#
NAT: s = 192.168.0.2 -> 16.16.1.1, d = 192.168.1.2 [29]

NAT*: s = 192.168.1.2, d = 16.16.1.1 -> 192.168.0.2 [50]

NAT: s = 192.168.0.2 -> 16.16.1.1, d = 192.168.1.2 [30]

NAT*: s = 192.168.1.2, d = 16.16.1.1 -> 192.168.0.2 [51]
Router0# show ip nat translations
Pro    Inside global      Inside local       Outside local      Outside global
icmp   16.16.1.1:1        192.168.0.3:1      192.168.1.2:1      192.168.1.2:1
```

//输出表明进行 PAT 转换可以使用同一个 IP 地址不同端口号。

【提示】 动态 NAT 的过期时间是 86 400s,PAT 的过期时间是 60s,通过"**show ip nat translations verbose**"命令可以查看。也可以通过下面的命令来修改超时时间:

Router0(config)# ip nat translation timeout *timeout* 参数 *timeout* 的范围是 0~2 147 483。

如果主机的数量不是很多,可以直接使用 outside 接口地址配置 PAT,不必定义地址池,命令如下:

Router0(config)# ip nat inside source list 2 interface s2/0 overload

第 6 章　动态路由协议配置

6.1　基础知识

6.1.1　RIP 概述

RIP 是由 Xerox 在 20 世纪 70 年代开发的,最初定义在 RFC1058 中。RIP 用两种数据包传输更新:更新和请求,每个有 RIP 功能的路由器默认情况下每隔 30s 利用 UDP 520 端口向与它直连的网络邻居广播(RIP v1)或组播(RIP v2)路由更新。因此路由器不知道网络的全局情况,如果路由更新在网络上传播慢,将会导致网络收敛较慢,造成路由环路。为了避免路由环路,RIP 采用水平分割、毒性逆转、定义最大跳数、闪式更新、抑制计时 5 个机制来避免路由环路。

RIP 有以下一些主要特性:

(1) RIP 属于典型的距离向量路由选择协议。

(2) RIP 消息通过广播地址 255.255.255.255 进行发送,使用 UDP 的 520 端口。

(3) RIP 以到目的网络的最小跳数作为路由选择度量标准,而不是在链路的带宽和延迟的基础上进行选择。

(4) RIP 是为小型网络设计的。它的跳数计数限制为 15 跳,16 跳为不可到达。

(5) RIP-1 是一种有类路由协议,不支持不连续子网设计。RIP-2 支持 CIDR 及 VLSM 可变长子网掩码,使其支持不连续子网设计。

(6) RIP 周期进行路由更新,将路由表广播给邻居路由器,广播周期默认为 30s。

(7) RIP 的管理距离为 120。

6.1.2　RIP 认证

RIP v1 不支持认证,如果设备配置 RIP v2 路由协议,可以在相应的接口配置认证,RIP 认证分为明文认证和 MD5 加密认证两种类型。这两种认证模式的工作方式完全相同,只是 MD5 发送的是消息摘要,而不是认证密钥本身。明文认证在线路上发送认证密钥本身。MD5 消息摘要是由密钥和消息产生的,而密钥本身并不发送,以防止在其传输过程中被窃取。需要交换路由更新的路由器之间的认证模式和密钥必须一致。

6.1.3　自动汇总

RIP v2 中,如果路由满足自动汇总的条件,就会被自动汇总为整个有类网络地址。基本自动汇总条件如下:如果某个端口上被公告的一条路由与端口的基于类的网络地址不同,那么被公告的网络的所有子网将作为一个表项(整个基于类的网络)被公告。

6.1.4 IGRP 概述

IGRP 是一种距离向量型的内部网关协议(IGP)。距离向量路由协议要求每个路由器以规则的时间间隔向其相邻的路由器发送其路由表的全部或部分。随着路由信息在网络上扩散,路由器就可以计算到所有节点的距离。

IGRP 使用一组 metric 的组合(向量),网络延迟、带宽、可靠性和负载都被用于路由选择,网管可以为每种 metric 设置权值,IGRP 可以用管理员设置的或默认的权值来自动计算最佳路由。

6.1.5 IGRP 的度量权重

IGRP 使用一个复合度量标准,它通过带宽、延迟、负载和可靠性来计算。默认情况下,只考虑带宽和延迟特性;其他参数值仅在配置启用后才考虑。延迟和带宽不是测试值,但是可以使用 delay 和 bandwidth 接口命令来设置。

度量标准=$[k1×带宽+(k2×带宽)/(256-负载)+k3×延迟]×[k5/(可靠性)+k4]$

度量标准 $k1$ 表示带宽,度量标准 $k3$ 表示延迟。默认情况下,度量标准 $k1$ 和 $k3$ 的值设置为1,而 $k2$、$k4$ 和 $k5$ 设置为0。

6.1.6 OSPF 概述

OSPF 是一个链路状态协议,其操作以网络连接或者链路的状态为基础。在 OSPF 中,计算网络拓扑时最基本的元素是每台路由器中每条链路的状态。通过学习每条链路连接到哪里,OSPF 可以建立一个数据库,记录网络中的所有链路,然后使用最短路径优先算法计算出到每个目标网络的最短路径。由于所有路由器都持有完全相同的网络拓扑图,所以 OSPF 不需要定时发送路由更新信息。OSPF 只有在网络拓扑发生变化的情况下才会发送路由更新信息。

OSPF 具有许多突出的优点:

(1) 开销减少。
(2) 支持 VLSM 和 CIDR。
(3) 支持不连续网络。
(4) 支持手工路由汇总。
(5) 收敛时间短。
(6) 无环拓扑生成。
(7) 跳步数只受路由器资源使用和 IP TTL 的限制。

6.1.7 OSPF 认证

OSPF 允许路由器之间互相进行身份验证。在默认的情况下,路由器相信从一台路由器发送过来的路由选择信息。路由器也相信这些信息没有被修改过。为了保证这种信任,在一定区域内的路由器可以配置成互相验证。

6.1.8 EIGRP 概述

EIGRP 路由协议属于一种混合型的路由协议,它在路由的学习方法上具有链路状态路由协议的特点,而它计算路径度量值的算法又具有距离矢量路由协议的特点。但是由于 EIGRP 路由协议是增强的 IGRP 路由协议,它是由 IGRP 路由协议发展而来的。EIGRP 是基于 IGRP 专有路由选择协议,所以只有 Cisco 的路由器之间可以使用该路由协议。

EIGRP 是一个高效的路由协议,它的特点如下:

(1) 通过发送和接收 Hello 包来建立和维持邻居关系,并交换路由信息。
(2) 采用组播(224.0.0.10)或单播进行路由更新。
(3) EIGRP 的管理距离为 90 或 170。
(4) 采用触发更新,减少带宽占用。
(5) 支持可变长子网掩码(VLSM),默认开启自动汇总功能。
(6) 支持 IP、IPX、AppleTalk 等多种网络层协议。
(7) 对每一种网络协议,EIGRP 都维持独立的邻居表、拓扑表和路由表。
(8) EIGRP 使用 Diffusing Update 算法(DUAL)来实现快速收敛,并确保没有路由环路。
(9) 存储整个网络拓扑结构的信息,以便快速适应网络变化。
(10) 支持等价和非等价的负载均衡。
(11) 使用可靠传输协议(RTP)保证路由信息传输的可靠性。
(12) 无缝连接数据链路层协议和拓扑结构,EIGRP 不要求对 OSI 参考模型的 2 层协议做特别的配置。

6.1.9 EIGRP 认证

为启用 EIGRP 认证,在接口配置模式下,使用 ip authentication key-chain eigrp 命令。为禁用该认证,使用该命令的 no 形式。

```
ip authentication key-chain eigrp as-number key-chain
no ip authentication key-chain eigrp as-number key-chain
```

6.2 实验 1 RIP v1 路由协议配置

6.2.1 实验拓扑

配置 RIP v1 动态路由协议,使得两台路由器模拟远程网络互联,见图 6.1。

设备名	IP 地址	端口
Router0	16.16.1.1	S2/0
Router0	19.168.0.1	F0/0
Router1	16.16.1.2	S2/0
Router1	19.168.1.1	F0/0
PC0	192.168.0.2	F0
PC1	192.168.1.2	F0

6.2.2 实验内容

掌握 RIP v1 动态路由协议的配置、诊断方法。
(1) 配置 RIP v1 动态路由协议。
(2) 验证配置信息。

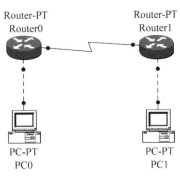

图 6.1 动态路由协议连接方式

6.2.3 实验步骤

(1) 配置 RIP。

```
Router0(config-if)#router rip
//默认版本为 version1.
Router0(config-router)#net 192.168.0.0
Router0(config-router)#net 16.16.1.0
```

```
Router1(config-if)#router rip
Router1(config-router)#net 192.168.1.0
Router1(config-router)#net 16.16.1.0
```

(2) 验证配置信息。

```
Router0#show ip protocols
Routing Protocol is "rip"
```

//应用 RIP。

```
Sending updates every 30 seconds, next due in 0 seconds
```

//更新周期是 30s,距离下次更新还有 0s。

```
Invalid after 180 seconds, hold down 180, flushed after 240
```

//路由条目如果在 180s 还没有收到更新,则被标记为无效。

```
Outgoing update filter list for all interfaces is not set
```

//在出方向上没有设置过滤列表。

```
Incoming update filter list for all interfaces is not set
Redistributing: rip
```

//只运行 RIP,没有其他的协议重分布进来。

```
Default version control: send version 1, receive any version
```

//默认发送版本 1 的路由更新,接收多种版本的路由更新。

```
  Interface          Send    Recv    Triggered RIP Key-chain
    FastEthernet0/0    1       2 1
    Serial2/0          1       2 1
```

//以上三行显示了运行 RIP 的接口,以及可以接收和发送的 RIP 路由更新的版本。

```
Automatic network summarization is in effect
```

//RIP 路由协议默认开启自动汇总功能。

```
Maximum path: 4
Routing for Networks:
    16.0.0.0
    192.168.0.0
```

//路由网络。

```
Passive Interface(s):
Routing Information Sources:
    Gateway         Distance        Last Update
    16.16.1.2       120             00:00:12
Distance: (default is 120)
```

//上述信息表明路由信息源,其中,gateway 为学习路由信息的路由器的接口地址,也就是下一跳地
//址;distance 为管理距离;last update 为更新发生在多长时间以前。

【提示】 可以通过下面的命令来调整上述三个时间参数。

```
Router0(config-router)#timers basic update invalid holddown flushed
router0#show ip route
Codes: C - connected, S - static, I - IGRP, R - RIP, M - mobile, B - BGP
       D - EIGRP, EX - EIGRP external, O - OSPF, IA - OSPF inter area
       N1 - OSPF NSSA external type 1, N2 - OSPF NSSA external type 2
       E1 - OSPF external type 1, E2 - OSPF external type 2, E - EGP
       i - IS-IS, L1 - IS-IS level-1, L2 - IS-IS level-2, ia - IS-IS inter area
       * - candidate default, U - per-user static route, o - ODR
       P - periodic downloaded static route

Gateway of last resort is not set

     16.0.0.0/24 is subnetted, 1 subnets
C       16.16.1.0 is directly connected, Serial2/0
C    192.168.0.0/24 is directly connected, FastEthernet0/0
R    192.168.1.0/24 [120/1] via 16.16.1.2, 00:00:00, Serial2/0
```

//输出表明路由器 Router0 学到了 1 条 RIP 路由,路由条目"R 192.168.1.0/24 [120/1] via 16.16.1.2,
//00:00:00,Serial2/0"的含义如下。
//① R:路由条目是通过 RIP 路由协议学习来的。
//② 192.168.1.0/24:目的网络。
//③ 120:RIP 路由协议的默认管理距离。
//④ 1:度量值,从路由器 Router0 到达网络 192.168.1.0/24 的度量值为 1 跳。
//⑤ 16.16.1.2:下一跳地址。
//⑥ 00:00:00:距离下一次更新还有 30(30~0)s。
//⑦ Serial2/0:接收该路由条目的本路由器的接口。
//同时通过该路由条目的掩码长度可以看到,RIP v1 确实不传递子网信息。

6.3 实验 2 RIP v2 路由协议配置

6.3.1 实验拓扑

配置 RIP v2 动态路由协议,使得多台路由器模拟远程网络互联,见图 6.2。

设备名	IP 地址	端口
Router0	16.16.1.1	S2/0
Router0	192.168.0.1	F0/0
Router1	16.16.1.2	S2/0
Router1	16.16.2.1	S3/0
Router2	16.16.2.2	S2/0
Router2	16.16.3.1	S3/0
Router3	16.16.3.2	S2/0
Router3	192.168.1.1	F0/0
PC0	192.168.0.2	F0
PC1	192.168.1.2	F0

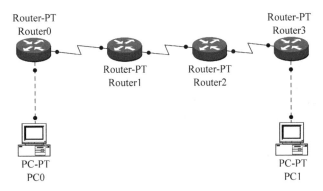

图 6.2 RIP v2 路由协议配置方式

6.3.2 实验内容

(1) 配置 RIP v2 动态路由协议。
(2) 启用参与路由协议的接口,并且通告网络。
(3) 验证 RIP v2 路由协议相关信息。

6.3.3 实验步骤

(1) 配置相关路由器。

```
Router0(config) # router rip
Router0(config - router) # v 2
Router0(config - router) # no auto - summary
Router0(config - router) # net 192.168.0.0
Router0(config - router) # net 16.16.1.0
```

```
Router1(config-router)#router rip
Router1(config-router)#v 2
Router1(config-router)#no auto-summary
Router1(config-router)#net 16.16.2.0
Router1(config-router)#net 16.16.1.0
```

```
Router2(config-if)#router rip
Router2(config-router)#v 2
Router2(config-router)#no auto-summary
Router2(config-router)#net 16.16.2.0
Router2(config-router)#net 16.16.3.0
```

```
Router3(config-if)#router rip
Router3(config-router)#v 2
Router3(config-router)#no auto-summary
Router3(config-router)#net 16.16.3.0
Router3(config-router)#net 192.168.1.0
```

（2）验证配置。

```
Router0#show ip route
Codes: C - connected, S - static, I - IGRP, R - RIP, M - mobile, B - BGP
       D - EIGRP, EX - EIGRP external, O - OSPF, IA - OSPF inter area
       N1 - OSPF NSSA external type 1, N2 - OSPF NSSA external type 2
       E1 - OSPF external type 1, E2 - OSPF external type 2, E - EGP
       i - IS-IS, L1 - IS-IS level-1, L2 - IS-IS level-2, ia - IS-IS inter area
       * - candidate default, U - per-user static route, o - ODR
       P - periodic downloaded static route

Gateway of last resort is not set

     16.0.0.0/24 is subnetted, 3 subnets
C       16.16.1.0 is directly connected, Serial2/0
R       16.16.2.0 [120/1] via 16.16.1.2, 00:00:06, Serial2/0
R       16.16.3.0 [120/2] via 16.16.1.2, 00:00:06, Serial2/0
C    192.168.0.0/24 is directly connected, FastEthernet0/0
R    192.168.1.0/24 [120/3] via 16.16.1.2, 00:00:06, Serial2/0
```

//上面输出的路由条目"192.168.1.0/24"，可以看到 RIP v2 路由更新是携带子网信息的。

```
Router0#show ip protocols
Routing Protocol is "rip"
Sending updates every 30 seconds, next due in 16 seconds
Invalid after 180 seconds, hold down 180, flushed after 240
Outgoing update filter list for all interfaces is not set
Incoming update filter list for all interfaces is not set
Redistributing: rip
Default version control: send version 2, receive 2
```

```
Interface          Send  Recv  Triggered RIP  Key-chain
Serial2/0           2     2
FastEthernet0/0     2     2
Automatic network summarization is not in effect
Maximum path: 4
Routing for Networks:
    16.0.0.0
    192.168.0.0
Passive Interface(s):
Routing Information Sources:
    Gateway        Distance        Last Update
    16.16.1.2      120             00:00:03
Distance: (default is 120)
```

//RIP v2 默认情况下只接收和发送版本 2 的路由更新。

【提示】 可以通过"ip rip send version"和"ip rip receive version"命令来控制在路由器接口上接收和发送的版本,例如,在 s2/0 接口上接收版本 1 和 2 的路由更新,但是只发送版本 2 的路由更新,配置如下:

```
Router0(config-if)#ip rip send version 2
Router0(config-if)#ip rip receive version 1 2
```

```
Router0#debug ip rip
RIP protocol debugging is on
Router0#RIP: received v2 update from 16.16.1.2 on Serial2/0

    16.16.2.0/24 via 0.0.0.0 in 1 hops

    16.16.3.0/24 via 0.0.0.0 in 2 hops

    192.168.1.0/24 via 0.0.0.0 in 3 hops

RIP: sending v2 update to 224.0.0.9 via Serial2/0 (16.16.1.1)

RIP: build update entries

    192.168.0.0/24 via 0.0.0.0, metric 1, tag 0

RIP: sending v2 update to 224.0.0.9 via FastEthernet0/0 (192.168.0.1)

RIP: build update entries

    16.16.1.0/24 via 0.0.0.0, metric 1, tag 0

    16.16.2.0/24 via 0.0.0.0, metric 2, tag 0

    16.16.3.0/24 via 0.0.0.0, metric 3, tag 0
```

```
        192.168.1.0/24 via 0.0.0.0, metric 4, tag 0


Router0#RIP: received v2 update from 16.16.1.2 on Serial2/0

    16.16.2.0/24 via 0.0.0.0 in 1 hops

    16.16.3.0/24 via 0.0.0.0 in 2 hops

    192.168.1.0/24 via 0.0.0.0 in 3 hops
```

【提示】 通过以上输出,可以看到 RIP v2 采用组播(224.0.0.9)更新,分别向 f0/0 和 s2/0 发送路由更新,同时从 s2/0 接收三条路由更新,分别是 192.168.1.0/24,度量值是 3 跳;16.16.3.0/24,度量值是 2 跳;16.16.2.0/24,度量值是 1 跳。

6.4 实验3 RIP v2 认证和触发更新

6.4.1 实验拓扑

配置 RIP v2 动态路由协议,使得多台路由器模拟远程网络互联,见图 6.3。

设备名	IP 地址	端口
Router0	16.16.1.1	S2/0
Router0	192.168.0.1	F0/0
Router1	16.16.1.2	S2/0
Router1	16.16.2.1	S3/0
Router2	16.16.2.2	S2/0
Router2	16.16.3.1	S3/0
Router3	16.16.3.2	S2/0
Router3	192.168.1.1	F0/0
PC0	192.168.0.2	F0
PC1	192.168.1.2	F0

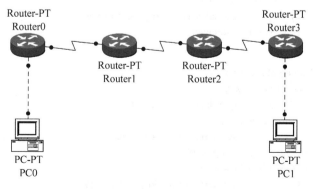

图 6.3 RIP v2 认证更新配置

6.4.2 实验内容

(1) RIP v2 明文认证的配置和匹配原则。
(2) RIP v2 MD5 认证的配置和匹配原则。
(3) RIP v2 触发更新。

6.4.3 实验步骤

(1) 配置明文认证。

```
Router0(config)#key chain test
```

//配置钥匙链。

```
Router0(config-keychain)#key 1
```

//配置 KEY ID。

```
Router0(config-keychain-key)#key-string cisco
```

//配置 KEY ID 的密匙。

```
Router0(config)#interface s2/0
Router0(config-if)#ip rip authentication mode text
```

//启用认证,认证模式为明文,默认认证模式就是明文,所以也可以不用指定。

```
Router0(config-if)#ip rip authentication key-chain test
```

//在接口上调用钥匙链。

```
Router0(config-if)#ip rip triggered
```

//在接口上启用触发更新。

```
Router1(config)#key chain test
Router1(config-keychain)#key 1
Router1(config-keychain-key)#key-string cisco
Router1(config)#interface s2/0
Router1(config-if)#ip rip triggered
Router1(config-if)#ip rip authentication key-chain test
Router1(config-if)#interface s3/0
Router1(config-if)#ip rip authentication key-chain test
Router1(config-if)#ip rip triggered
```

```
Router2(config)#key chain test
Router2(config-keychain)#key 1
Router2(config-keychain-key)#key-string cisco
Router2(config)#interface s2/0
Router2(config-if)#ip rip authentication key-chain test
Router2(config-if)#ip rip triggered
Router2(config-if)#interface s3/0
Router2(config-if)#ip rip authentication key-chain test
Router2(config-if)#ip rip triggered
```

```
Router3(config)#key chain test
Router3(config-keychain)#key 1
Router3(config-keychain-key)#key-string cisco
Router3(config)#interface s2/0
Router3(config-if)#ip rip authentication key-chain test
Router3(config-if)#ip rip triggered
```

(2) 验证调试。

```
Router1#show ip protocols
Routing Protocol is "rip"
Outgoing update filter list for all interfaces is not set
Incoming update filter list for all interfaces is not set
Sending updates every 30 seconds, next due in 4 seconds
Invalid after 180 seconds, hold down 0, flushed after 240
```

//由于触发更新,hold down 计时器自动为 0。

```
Redistributing: rip
Default version control: send version 2, receive version 2
  Interface Send Recv Triggered RIP Key-chain
  Serial2/0  2    2    Yes           test
  Serial3/0  2    2    Yes           test
```

//以上两行表明 s2/0 和 s3/0 接口启用了认证和触发更新。

```
Automatic network summarization is not in effect
Maximum path: 4
Routing for Networks:
  16.0.0.0
Routing Information Sources:
  Gateway      Distance    Last Update
  16.16.1.1    120         00:26:10
  16.16.2.2    120         00:26:01
Distance: (default is 120)
```

```
Router1#debug ip rip
RIP protocol debugging is on
192.168.10/24 via 0.0.0.0 in 2 hops
RIP: received packet with text authentication cisco
RIP: received v2 triggered update from 16.16.2.2 on Serial3/0
RIP: sending v2 ack to 16.16.2.2 via Serial3/0(16.16.2.1),flush, seq# 3
16.16.3.0/24 via 0.0.0.0 in 1 hops
```

//从上面的输出可以看出,在路由器 Router1 上,虽然打开了 debug ip rip,但是由于采用触发更新,所
//以并没有看到每 30s 更新一次的信息,而是清除了路由表这件事件触发了路由更新。而且所有的
//更新中都有"triggered"的字样,同时在接收的更新中带有"text authentication"的字样,证明接口 s2/0
//和 s3/0 启用了触发更新和明文认证。

6.5 实验 4 EIGRP 基本配置

6.5.1 实验拓扑

根据提供路由器,配置基本 EIGRP 连通网络,见图 6.4。

设备名	IP 地址	端口
Router0	192.168.0.1	F0/0
Router0	16.16.1.1	S2/0
Router1	192.168.1.1	F0/0
Router1	16.16.1.2	S2/0
PC0	192.168.0.2	F0
PC1	192.168.1.2	F0

6.5.2 实验内容

(1) 掌握 EIGRP 路由协议的配置方法。
(2) 掌握 EIGRP 度量值的计算方法。
(3) 理解邻居表、拓扑表以及路由表的含义。
(4) 验证调试。

6.5.3 实验步骤

(1) 路由器配置。

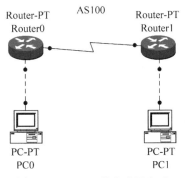

图 6.4 EIGRP 基本配置方式

```
Router0(config)# router eigrp 100
```

//开启 EIGRP 路由协议。

```
Router0(config-router)# net 16.16.1.0
Router0(config-router)# net 192.168.0.0
```

```
Router1(config)# router eigrp 100
Router1(config-router)# net 16.16.1.0
Router1(config-router)# net 192.168.1.0
```

【提示】 EIGRP 在通告网段时,如果是主类网络(即标准 A、B、C 类的网络,或者说没有划分子网的网络),则只需输入此网络地址;如果是子网,则最好在网络号后面写子网掩码或者反掩码,这样可以避免将所有的子网都加入 EIGRP 进程中。

反掩码是用广播地址(255.255.255.255)减去子网掩码所得到的。如掩码地址是 255.255.248.0,则反掩码地址是 0.0.7.255。在高级的 IOS 中也支持网络掩码的写法。运行 EIGRP 的整个网络 AS 号码必须一致,其范围为 1~65 535 之间。

(2) 验证调试。

```
Router0 # show ip route
Codes: C - connected, S - static, I - IGRP, R - RIP, M - mobile, B - BGP
       D - EIGRP, EX - EIGRP external, O - OSPF, IA - OSPF inter area
       N1 - OSPF NSSA external type 1, N2 - OSPF NSSA external type 2
       E1 - OSPF external type 1, E2 - OSPF external type 2, E - EGP
       i - IS-IS, L1 - IS-IS level-1, L2 - IS-IS level-2, ia - IS-IS inter area
       * - candidate default, U - per-user static route, o - ODR
       P - periodic downloaded static route

Gateway of last resort is not set

     16.0.0.0/8 is variably subnetted, 2 subnets, 2 masks
D       16.0.0.0/8 is a summary, 00:04:41, Null0
C       16.16.1.0/24 is directly connected, Serial2/0
C    192.168.0.0/24 is directly connected, FastEthernet0/0
D    192.168.1.0/24 [90/20514560] via 16.16.1.2, 00:04:04, Serial2/0
```

//输出表明路由器 Router0 通过 EIGRP 学到了两条 EIGRP 路由条目,管理距离是 90。

① 注意 EIGRP 代码用字母"D"表示,如果通过重分布方式进入 EIGRP 网络的路由条目,默认管理距离为 170,路由代码用"D EX"表示,也说明 EIGRP 路由协议能够区分内部路由和外部路由。

② 对于 EIGRP 度量值的计算,不妨以"D 192.168.1.0/24 [90/20514560] via 16.16.1.2, 00:04:04,Serial2/0"路由条目为例来说明。

EIGRP 度量值的计算公式 $=[K1\times Bandwidth+(K2\times Bandwidth)/(256-Load)+K3\times Delay]\times[K5/(Reliability+K4)]\times 256$

默认情况下,$K1=K3=1,K2=K4=K5=0$。

Bandwidth $=10^7$/所经由链路中入口带宽(单位为 kb/s)的最小值

Delay=所经由链路中入口的延迟之和(单位为 μs)/10

接下来看一下在路由器 Router0 中的"192.168.1.0"路由条目的度量值是如何计算的。

首先看带宽应该是从 Router1 的 f0/0 口到 Router0 最小的,应该是 Router0 的 s2/0 接口的带宽,为 128k,而延迟是路由器 Router1 的 192.168.1.1 和路由器 Router0 的 s2/0 接口的延迟之和,所以最后的度量值应该是 $[10^7/128+(100+20\,000)/10]\times 256=20\,514\,560$,和路由器计算的结果是一致的。

【提示】 接口的带宽和延迟可以通过"show interface"查看。

```
Router0 # show ip protocols

Routing Protocol is "eigrp 100 "
```

//as 号为 100。

```
  Outgoing update filter list for all interfaces is not set
  Incoming update filter list for all interfaces is not set
```

```
    Default networks flagged in outgoing updates
    Default networks accepted from incoming updates
EIGRP metric weight K1 = 1, K2 = 0, K3 = 1, K4 = 0, K5 = 0
```

//计算度量值中参数值。

```
EIGRP maximum hopcount 100
```

//最大跳数。

```
    EIGRP maximum metric variance 1
Redistributing: eigrp 100
Automatic network summarization is in effect
```

//显示自动汇总开启,默认自动汇总是开启的。

```
  Automatic address summarization:
    16.0.0.0/8 for FastEthernet0/0
      Summarizing with metric 20512000
Maximum path: 4
Routing for Networks:
    16.0.0.0
    192.168.0.0
Routing Information Sources:
    Gateway       Distance      Last Update
    16.16.1.2     90            3602464
Distance: internal 90 external 170
```

```
Router0#show ip eigrp neighbors
IP-EIGRP neighbors for process 100
H  Address      Interface     Hold Uptime    SRTT    RTO     Q      Seq
                              (sec)          (ms)            Cnt    Num
0  16.16.1.2    Se2/0         13 00:36:46    40      1000    0      4
```

输出各字段的含义如下。

① H:表示与邻居建立会话的顺序。

② Address:邻居路由器的接口地址。

③ Interface:本地到邻居路由器的接口。

④ Hold:认为邻居关系不存在所能等待的最大时间。

⑤ Uptime:从邻居关系建立到目前的时间。

⑥ SRTT:向邻居路由器发送一个数据包以及本路由器收到确认包的时间。

⑦ RTO:路由器在重新传输包之前等待 ACK 的时间。

⑧ Q Cnt:等待发送的队列。

⑨ Seq Num:从邻居收到的发送数据包的序列号。

【提示】 运行 EIGRP 路由协议的路由器不能建立邻居关系的可能原因:①EIGRP 进程的 AS 号码不同;②计算度量值的 K 值不同。

```
Router0# show ip eigrp interfaces
IP-EIGRP interfaces for process 100

                    Xmit Queue    Mean   Pacing Time   Multicast    Pending
Interface   Peers   Un/Reliable   SRTT   Un/Reliable   Flow Timer   Routes
Se2/0       1       0/0           1236   0/10          0            0
Fa0/0       0       0/0           1236   0/10          0            0
```

输出各字段的含义如下。

① Interface：运行 EIGRP 的接口。

② Peers：该接口的邻居的个数。

③ Xmit Queue Un/Reliable：在不可靠/可靠队列中存留的数据包的数量。

④ Mean SRTT：平均的往返时间，单位是 s。

⑤ Pacing Time Un/Reliable：用来确定不可靠/可靠队列中数据包被送出接口的时间间隔。

⑥ Multicast Flow Timer：组播数据包被发送前最长的等待时间。

⑦ Pending Routes：在传送队列中等待被发送的数据包携带的路由条目。

```
Router0# show ip eigrp traffic
IP-EIGRP Traffic Statistics for process 100
  Hellos sent/received: 1095/541
  Updates sent/received: 2/3
  Queries sent/received: 0/0
  Replies sent/received: 0/0
  Acks sent/received: 3/2
  Input queue high water mark 1, 0 drops
  SIA-Queries sent/received: 0/0
SIA-Replies sent/received: 0/0
```

//输出显示了 EIGRP 发送和接收到的数据包的统计情况。

【提示】 在 EIGRP 中，有 5 种类型的数据包。

① Hello：以组播的方式定期发送，用于建立和维持邻居关系。

② 更新：当路由器收到某个邻居路由器的第一个 Hello 包时，以单播传送方式回送一个包含它所知道的路由信息的更新包。当路由信息发生变化时，以组播的方式发送只包含变化信息的更新包。

③ 查询：当一条链路失效，路由器重新进行路由计算，但在拓扑表中没有可行的后继路由时，路由器就以组播的方式向它的邻居发送一个查询包，以询问它们是否有一条到目的地的后继路由。

④ 答复：以单播的方式回传给查询方，对查询数据包进行应答。

⑤ 确认：以单播的方式传送，用来确认更新、查询、答复数据包。

6.6 实验 5 EIGRP 路由汇总

6.6.1 实验拓扑

根据提供路由器,配置 EIGRP,验证路由汇总功能,见图 6.5。

设备名	IP 地址	端口
Router0	16.16.1.1	S2/0
Router1	16.16.1.2	S2/0
Router1	16.16.2.1	S3/0
Router2	16.16.2.2	S2/0
Router2	16.16.3.1	S3/0
Router3	16.16.3.2	S2/0
Router3	4.4.0.4	Loopback0
Router3	4.4.1.4	Loopback1
Router3	4.4.1.4	Loopback2
Router3	4.4.1.4	Loopback3

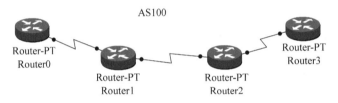

图 6.5 EIGRP 路由汇总连接方式

6.6.2 实验内容

(1) 掌握路由汇总的目的。
(2) 掌握 EIGRP 自动汇总。
(3) 掌握 EIGRP 手工汇总。
(4) 理解指向 null0 路由的含义。
(5) 验证调试。

6.6.3 实验步骤

(1) 路由器配置。

```
Router0(config-if)#router eigrp 100
Router0(config-router)#no auto-summary
```

//关闭自动汇总功能。

```
Router0(config-router)#net 16.16.1.0 255.255.255.0
```

```
Router1(config-if)#router eigrp 100
Router1(config-router)#no auto-summary
Router1(config-router)#net 16.16.1.0 255.255.255.0
Router1(config-router)#net 16.16.2.0
```

```
Router2(config)#router eigrp 100
Router2(config-router)#no auto-summary
Router2(config-router)#net 16.16.2.0
Router2(config-router)#net 16.16.3.0
```

```
Router3(config)#router eigrp 100
Router3(config-router)#no auto-summary
Router3(config-router)#net 4.4.0.0 255.255.252.0
Router3(config-router)#net 16.16.3.0
Router3(config-router)#int s2/0
Router3(config-if)#ip summary-address eigrp 100 4.4.0.0 255.255.252.0
```

//配置 EIGRP 手工路由汇总。

(2) 验证调试。

```
Router2#show ip route eigrp
     4.0.0.0/22 is subnetted, 1 subnets
D       4.4.0.0 [90/20640000] via 16.16.3.2, 00:03:57, Serial3/0
     16.0.0.0/24 is subnetted, 3 subnets
D       16.16.1.0 [90/21024000] via 16.16.2.1, 00:06:31, Serial2/0
```

//输出表明路由器 Router2 的路由表中有 4 条明细路由。

```
Router3#show ip route eigrp
     4.0.0.0/8 is variably subnetted, 5 subnets, 2 masks
D       4.4.0.0/22 is a summary, 00:08:08, Null0
     16.0.0.0/24 is subnetted, 3 subnets
D       16.16.1.0 [90/21536000] via 16.16.3.1, 00:08:07, Serial2/0
D       16.16.2.0 [90/21024000] via 16.16.3.1, 00:08:07, Serial2/0
```

【提示】 当被汇总的明细路由全部 down 掉以后,汇总路由才自动从路由表里被删除,从而可以有效地避免路由抖动。

6.7 实验6 EIGRP 认证

6.7.1 实验拓扑

根据提供路由器,配置 EIGRP 认证,验证配置,见图 6.6。

设备名	IP 地址	端口
Router0	16.16.1.1	S2/0
Router1	16.16.1.2	S2/0
Router1	16.16.2.1	S3/0
Router2	16.16.2.2	S2/0
Router2	16.16.3.1	S3/0
Router3	16.16.3.2	S2/0
Router3	4.4.0.4	Loopback0
Router3	4.4.1.4	Loopback1
Router3	4.4.1.4	Loopback2
Router3	4.4.1.4	Loopback3

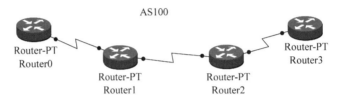

图 6.6　EIGRP 认证连接方式

6.7.2　实验内容

（1）掌握 EIGRP 路由协议认证的配置。

（2）验证调试。

6.7.3　实验步骤

（1）配置路由器。

```
Router0(config) # key chain ccnp
Router0(config - keychain) # key 1
Router0(config - keychain - key) # key - string cisco
Router0(config) # interface s0/0/0
Router0(config - if) # ip authentication mode eigrp 1 md5
```

//认证模式为 MD5。

```
Router1 (config - if) # ip authentication key - chain eigrp 1 ccnp
```

//在接口上调用钥匙链。

```
Router1(config) # key chain ccnp
Router1(config - keychain) # key 1
Router1(config - keychain - key) # key - string cisco
Router1(config) # interface s0/0/0
Router1(config - if) # ip authentication mode eigrp 1 md5
Router1(config - if) # ip authentication key - chain eigrp 1 ccnp
Router1(config) # interface s0/0/1
Router1(config - if) # ip authentication mode eigrp 1 md5
Router1(config - if) # ip authentication key - chain eigrp 1 ccnp
```

```
Router2(config)#key chain ccnp
Router2(config-keychain)# key 1
Router2(config-keychain-key)#key-string cisco
Router2(config)#interface s0/0/0
Router2(config-if)#ip authentication mode eigrp 1 md5
Router2(config-if)#ip authentication key-chain eigrp 1 ccnp
Router2(config)#interface s0/0/1
Router2(config-if)#ip authentication mode eigrp 1 md5
Router2(config-if)#ip authentication key-chain eigrp 1 ccnp
```

```
Router3(config)#key chain ccnp
Router3(config-keychain)# key 1
Router3(config-keychain-key)#key-string cisco
Router3(config)#interface s0/0/0
Router3(config-if)#ip authentication mode eigrp 1 md5
Router3(config-if)#ip authentication key-chain eigrp 1 ccnp
```

(2) 验证调试。

① 如果链路的一端启用了认证,另外一端没有起用认证,则出现下面的提示信息:

```
IP-EIGRP(0) 1: Neighbor 16.16.1.2(Serial2/0) is down: authentication mode changed
```

② 如果钥匙链的密匙不正确,则出现下面的提示信息:

```
IP-EIGRP(0) 1: Neighbor 16.16.1.1(Serial2/0) is down: Auth failure
```

6.8 实验 7 EIGRP 基本配置

6.8.1 实验拓扑

实验中路由器 EIGRP 配置方式见图 6.7。

设备名	IP 地址	端口
Router0	16.16.1.1	S2/0
Router0	19.168.0.1	F0/0
Router1	16.16.1.2	S2/0
Router1	16.16.2.1	S3/0
Router2	16.16.2.2	S2/0
Router2	16.16.3.1	S3/0
Router3	16.16.3.2	S2/0
Router3	19.168.1.1	F0/0
PC0	192.168.0.2	F0
PC1	192.168.1.2	F0

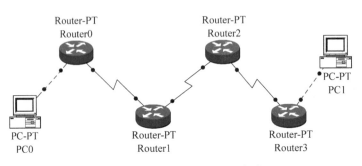

图 6.7 路由器 EIGRP 配置方式

6.8.2 实验内容

掌握 EIGRP 动态路由协议的配置、诊断方法。

(1) 配置 EIGRP 动态路由协议。
(2) 理解 EIGRP 度量值的计算方法。
(3) 查看并验证配置信息。

6.8.3 实验步骤

(1) 配置相关路由器。

```
Router0(config)# router eigrp 1
Router0(config-router)# no auto-summary
Router0(config-router)# net 192.168.0.0
Router0(config-router)# net 16.16.1.0 0.0.0.255
```

```
Router1(config)# router eigrp 1
Router1(config-router)# no auto-summary
Router1(config-router)# net 16.16.1.0
% DUAL-5-NBRCHANGE: IP-EIGRP 1: Neighbor 16.16.1.1 (Serial2/0) is up: new adjacency
Router1(config-router)# net 16.16.2.0
```

```
Router2(config)# router eigrp 1
Router2(config-router)# no auto-summary
Router2(config-router)# net 16.16.2.0
% DUAL-5-NBRCHANGE: IP-EIGRP 1: Neighbor 16.16.2.1 (Serial2/0) is up: new adjacency
Router2(config-router)# net 16.16.3.0
```

```
Router3(config)# router eigrp 1
Router3(config-router)# no auto-summary
Router3(config-router)# net 16.16.3.0
% DUAL-5-NBRCHANGE: IP-EIGRP 1: Neighbor 16.16.3.1 (Serial2/0) is up: new adjacency
Router3(config-router)# net 192.168.1.0
```

【提示】 EIGRP 在通告网段时,如果是主类网络(即标准 A、B、C 类的网络,或者说没有划分子网的网络),则只需输入此网络地址;如果是子网,则最好在网络号后面写子网掩码或者反掩码,这样可以避免将所有的子网都加入 EIGRP 进程中。

反掩码是用广播地址(255.255.255.255)减去子网掩码所得到的。如掩码地址是 255.255.248.0,则反掩码地址是 0.0.7.255。在高级的 IOS 中也支持网络掩码的写法。

运行 EIGRP 的整个网络 AS 号码必须一致,其范围为 1~65 535 之间。

(2) 验证调试。

```
Router1#show ip route
Codes: C - connected, S - static, I - IGRP, R - RIP, M - mobile, B - BGP
       D - EIGRP, EX - EIGRP external, O - OSPF, IA - OSPF inter area
       N1 - OSPF NSSA external type 1, N2 - OSPF NSSA external type 2
       E1 - OSPF external type 1, E2 - OSPF external type 2, E - EGP
       i - IS-IS, L1 - IS-IS level-1, L2 - IS-IS level-2, ia - IS-IS inter area
       * - candidate default, U - per-user static route, o - ODR
       P - periodic downloaded static route

Gateway of last resort is not set

     16.0.0.0/24 is subnetted, 3 subnets
C       16.16.1.0 is directly connected, Serial2/0
C       16.16.2.0 is directly connected, Serial3/0
D       16.16.3.0 [90/21024000] via 16.16.2.2, 00:06:45, Serial3/0
D    192.168.0.0/24 [90/20514560] via 16.16.1.1, 00:08:25, Serial2/0
D    192.168.1.0/24 [90/21026560] via 16.16.2.2, 00:05:25, Serial3/0
```

//以上输出表明路由器 Router1 通过 EIGRP 学到了三条 EIGRP 路由条目,管理距离是 90,注意 EIGRP
//代码用字母"D"表示,如果通过重分布方式进入 EIGRP 网络的路由条目,默认管理距离为 170,路
//由代码用"D EX"表示,也说明 EIGRP 路由协议能够区分内部路由和外部路由。

对于 EIGRP 度量值的计算,不妨以"D 192.168.0.0/24 [90/20514560] via 16.16.1.1, 00:08:25, Serial2/0"路由条目为例来说明。

EIGRP 度量值的计算公式=[$K1\times$Bandwidth + ($K2\times$Bandwidth)/(256-Load) +$K3\times$Delay]$\times$[$K5$/(Reliability + $K4$)]$\times$256

默认情况下,$K1=K3=1,K2=K4=K5=0$。

Bandwidth =10^7/所经由链路中入口带宽(单位为 kb/s)的最小值

Delay=所经由链路中入口的延迟之和(单位为 μs)/10

接下来看一下在路由器 Router1 中的"192.168.0.0/24"路由条目的度量值是如何计算的。

首先看带宽应该是从 Router0 的 192.168.0.0/24 到 Router1 最小的,应该是 Router1 的 S2/0 接口的带宽,为 128k,而延迟是路由器 Router0 的 192.168.0.0/24 和路由器 Router1 的 s2/0 接口的延迟之和,所以最后的度量值应该是[10^7/128+(100+20000)/10]$\times$256=20 514 560,和路由器计算的结果是一致的。

```
Router1#show ip protocols

Routing Protocol is "eigrp 1"
```

//AS 号码为 1。

```
Outgoing update filter list for all interfaces is not set
Incoming update filter list for all interfaces is not set
Default networks flagged in outgoing updates
Default networks accepted from incoming updates
EIGRP metric weight K1 = 1, K2 = 0, K3 = 1, K4 = 0, K5 = 0
```

//显示计算度量值所用的 K 值。

```
  EIGRP maximum hopcount 100
  EIGRP maximum metric variance 1
Redistributing: eigrp 1
  Automatic network summarization is not in effect
```

//显示自动汇总已经关闭,默认自动汇总是开启的。

```
Maximum path: 4
Routing for Networks:
   16.0.0.0
Routing Information Sources:
   Gateway        Distance       Last Update
   16.16.1.1      90             6441
   16.16.2.2      90             8080
Distance: internal 90 external 170
```

```
Router1#show ip eigrp neighbors
IP-EIGRP neighbors for process 1
H  Address     Interface   Hold Uptime    SRTT   RTO    Q     Seq
                           (sec)          (ms)          Cnt   Num
0  16.16.1.1   Se2/0       10   00:26:14  40     1000   0     7
1  16.16.2.2   Se3/0       11   00:26:12  40     1000   0     7
```

//以上输出各字段的含义如下:

//① H:与邻居建立会话的顺序。

//② Address:邻居路由器的接口地址。

//③ Interface:本地到邻居路由器的接口。

//④ Hold:认为邻居关系不存在所能等待的最大时间。

//⑤ Uptime:从邻居关系建立到目前的时间。

//⑥ SRTT:是向邻居路由器发送一个数据包以及本路由器收到确认包的时间。

//⑦ RTO:路由器在重新传输包之前等待 ACK 的时间。

//⑧ Q Cnt:等待发送的队列。

//⑨ Seq Num:从邻居收到的发送数据包的序列号。

```
Router1#show ip eigrp topology
IP-EIGRP Topology Table for AS 1

Codes: P - Passive, A - Active, U - Update, Q - Query, R - Reply,
       r - Reply status
```

```
P 16.16.2.0/24, 1 successors, FD is 20512000
        via Connected, Serial3/0
P 16.16.1.0/24, 1 successors, FD is 20512000
        via Connected, Serial2/0
P 192.168.0.0/24, 1 successors, FD is 20514560
        via 16.16.1.1 (20514560/28160), Serial2/0
P 16.16.3.0/24, 1 successors, FD is 21024000
        via 16.16.2.2 (21024000/20512000), Serial3/0
P 192.168.1.0/24, 1 successors, FD is 21026560
        via 16.16.2.2 (21026560/20514560), Serial3/0
```

//以上输出可以清楚地看到每条路由条目的 FD 和 RD 的值。而拓扑结构数据库中状态代码最常见
//的是"P"、"A"和"s",含义如下。
//① P：代表 passive,表示网络处于收敛的稳定状态。
//② A：代表 active,当前网络不可用,正处于发送查询状态。
//③ s：在 3min 内,如果被查询的路由没有收到回应,查询的路由就被置为 stuck inactive 状态。

【提示】 可行距离(FD)：到达一个目的网络的最小度量值。

通告距离(RD)：邻居路由器所通告的它自己到达目的网络的最小的度量值。

可行性条件(FC)：是 EIGRP 路由器更新路由表和拓扑表的依据。可行性条件可以有效地阻止路由环路,实现路由的快速收敛。可行性条件的公式为：AD<FD。

```
Router1#show ip eigrp interfaces
IP-EIGRP interfaces for process 1

            Xmit Queue    Mean    Pacing Time   Multicast     Pending
Interface   Peers  Un/Reliable  SRTT  Un/Reliable   Flow Timer   Routes
Se3/0       1      0/0          1236  0/10          0            0
Se2/0       1      0/0          1236  0/10          0            0
```

//以上输出各字段的含义如下。
//① Interface：运行 EIGRP 的接口。
//② Peers：该接口的邻居的个数。
//③ Xmit Queue Un/Reliable：在不可靠/可靠队列中存留的数据包的数量。
//④ Mean SRTT：平均的往返时间,单位是 s。
//⑤ Pacing Time Un/Reliable：用来确定不可靠/可靠队列中数据包被送出接口的时间间隔。
//⑥ Multicast Flow Timer：组播数据包被发送前最长的等待时间。
//⑦ Pending Routes：在传送队列中等待被发送的数据包携带的路由条目。

```
Router1#show ip eigrp traffic
IP-EIGRP Traffic Statistics for process 1
  Hellos sent/received: 908/906
  Updates sent/received: 7/8
  Queries sent/received: 0/0
  Replies sent/received: 0/0
  Acks sent/received: 8/6
  Input queue high water mark 1, 0 drops
  SIA-Queries sent/received: 0/0
  SIA-Replies sent/received: 0/0
```

6.9 实验 8 点到点链路 OSPF 配置

6.9.1 实验拓扑

根据提供路由器,配置点到点链路上的 OSPF,对运行中的 OSPF 进行诊断,见图 6.8。

设备名	IP 地址	端口
Router0	16.16.1.1	S2/0
Router0	19.168.0.1	F0/0
Router1	16.16.1.2	S2/0
Router1	16.16.2.1	S3/0
Router1	19.168.1.1	F0/0
Router2	16.16.2.2	S2/0
Router2	16.16.3.1	S3/0
Router2	19.168.2.1	F0/0
Router3	16.16.3.2	S2/0
Router3	19.168.2.1	F0/0
PC0	192.168.0.2	F0
PC1	192.168.1.2	F0
PC2	192.168.2.2	F0
PC3	192.168.3.2	F0

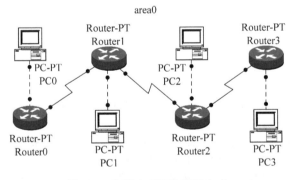

图 6.8 点到点 OSPF 配置方式

6.9.2 实验内容

(1) 在路由器上启动 OSPF 路由进程。
(2) 启用参与路由协议的接口,并且通告网络及所在的区域。
(3) 度量值 cost 的计算。
(4) hello 相关参数的配置。
(5) 点到点链路上的 OSPF 的特征。
(6) 查看和调试 OSPF 路由协议相关信息。

6.9.3 实验步骤

(1) 配置路由器 OSPF 协议。

```
Router0(config)#router ospf 1
Router0(config-router)#router
Router0(config-router)#router-id 192.168.0.1
Router0(config-router)#net 192.168.0.0 255.255.255.0 area 0
Router0(config-router)#net 16.16.1.0 255.255.255.0 area 0
```

```
Router1(config)#router ospf 1
Router1(config-router)#router-id 192.168.1.1
Router1(config-router)#net 192.168.1.0 255.255.255.0 area 0
Router1(config-router)#net 16.16.1.0 255.255.255.0 area 0
Router1(config-router)#net 16.16.2.0 255.255.255.0 area 0
```

```
Router2(config)#router ospf 1
Router2(config-router)#router-id 192.168.2.1
Router2(config-router)#net 192.168.2.0 255.255.255.0 area 0
Router2(config-router)#net 16.16.2.0 255.255.255.0 area 0
Router2(config-router)#net 16.16.3.0 255.255.255.0 area 0
```

```
Router3(config)#router ospf 1
Router3(config-router)#router-id 192.168.3.1
Router3(config-router)#net 192.168.3.0 255.255.255.0 area 0
Router3(config-router)#net 16.16.3.0 255.255.255.0 area 0
```

【提示】

① OSPF 路由进程 ID 的范围必须在 1~65 535 之间,而且只有本地含义,不同路由器的路由进程 ID 可以不同。如果要想启动 OSPF 路由进程,至少确保有一个接口是 up 的。

② 区域 ID 是在 0~4 294 967 295 内的十进制数,也可以是 IP 地址的格式 A.B.C.D。当网络区域 ID 为 0 或 0.0.0.0 时称为主干区域。

③ 在高版本的 IOS 中通告 OSPF 网络的时候,网络号的后面可以跟网络掩码,也可以跟反掩码,都是可以的。

④ 确定 Router ID 遵循如下顺序。

- 最优先的是在 OSPF 进程中用"router-id"命令指定了路由器 ID。
- 如果没有在 OSPF 进程中指定路由器 ID,那么选择 IP 地址最大的环回接口的 IP 地址为 Router ID。
- 如果没有环回接口,就选择最大的活动的物理接口的 IP 地址为 Router ID。

建议用"router-id"命令来指定路由器 ID,这样可控性比较好。

(2) 验证调试。

```
Router1#show ip route
Codes: C - connected, S - static, I - IGRP, R - RIP, M - mobile, B - BGP
```

```
       D - EIGRP, EX - EIGRP external, O - OSPF, IA - OSPF inter area
       N1 - OSPF NSSA external type 1, N2 - OSPF NSSA external type 2
       E1 - OSPF external type 1, E2 - OSPF external type 2, E - EGP
       i - IS-IS, L1 - IS-IS level-1, L2 - IS-IS level-2, ia - IS-IS inter area
       * - candidate default, U - per-user static route, o - ODR
       P - periodic downloaded static route

Gateway of last resort is not set

     16.0.0.0/24 is subnetted, 3 subnets
C       16.16.1.0 is directly connected, Serial2/0
C       16.16.2.0 is directly connected, Serial3/0
O       16.16.3.0 [110/1562] via 16.16.2.2, 00:11:09, Serial3/0
O    192.168.0.0/24 [110/782] via 16.16.1.1, 00:12:45, Serial2/0
C    192.168.1.0/24 is directly connected, FastEthernet0/0
O    192.168.2.0/24 [110/782] via 16.16.2.2, 00:11:09, Serial3/0
O    192.168.3.0/24 [110/1563] via 16.16.2.2, 00:09:57, Serial3/0
```

//同一个区域内通过 OSPF 路由协议学习的路由条目用代码"O"表示。

【提示】 路由条目"192.168.3.0"的度量值为1563,计算过程如下。

cost 的计算公式为 10^8/带宽(b/s),然后取整,而且是所有链路入口的 cost 之和,接口 f0/0 的 cost 为 1,路由条目"192.168.3.0"到路由器 Router1 经过的入接口包括路由器 router3 的 f0/0,路由器 Router2 的 s2/0,路由器 Router1 的 s3/0,所以计算如下:

$1+10^8/128\,000+10^8/128\,000=1563$。也可以直接通过"ip ospf cost"命令设置接口的 cost 值,并且它是优先计算 cost 值的。

```
Router1#show ip protocols

Routing Protocol is "ospf 1"
```

//路由器运行的 OSPF 进程 ID。

```
  Outgoing update filter list for all interfaces is not set
  Incoming update filter list for all interfaces is not set
Router ID 192.168.1.1
```

//本路由器 ID。

```
Number of areas in this router is 1. 1 normal 0 stub 0 nssa
```

//本路由器参与的区域数量和类型。

```
Maximum path: 4
```

//支持等价路径最大数目。

```
  Routing for Networks:
    192.168.1.0 0.0.0.255 area 0
    16.16.1.0 0.0.0.255 area 0
16.16.2.0 0.0.0.255 area 0
```

//OSPF 通告的网络以及这些网络所在的区域。

```
Routing Information Sources:
    Gateway         Distance        Last Update
    16.16.1.1       110             00:18:09
    16.16.2.2       110             00:18:12
```

//路由信息来源。

```
Distance: (default is 110)
```

//OSPF 路由协议默认的管理距离。

```
Router1#show ip ospf 1
Routing Process "ospf 1" with ID 192.168.1.1
Supports only single TOS(TOS0) routes
Supports opaque LSA
SPF schedule delay 5 secs, Hold time between two SPFs 10 secs
Minimum LSA interval 5 secs. Minimum LSA arrival 1 secs
Number of external LSA 0. Checksum Sum 0x000000
Number of opaque AS LSA 0. Checksum Sum 0x000000
Number of DCbitless external and opaque AS LSA 0
Number of DoNotAge external and opaque AS LSA 0
Number of areas in this router is 1. 1 normal 0 stub 0 nssa
External flood list length 0
    Area BACKBONE(0)
        Number of interfaces in this area is 3
        Area has no authentication
        SPF algorithm executed 5 times
        Area ranges are
        Number of LSA 4. Checksum Sum 0x01ebad
        Number of opaque link LSA 0. Checksum Sum 0x000000
        Number of DCbitless LSA 0
        Number of indication LSA 0
        Number of DoNotAge LSA 0
        Flood list length 0
Router1#show ip ospf interface s2/0
Serial2/0 is up, line protocol is up
Internet address is 16.16.1.2/24, Area 0
```

//该接口的地址和运行的 OSPF 区域。

```
Process ID 1, Router ID 192.168.1.1, Network Type POINT－TO－POINT, Cost: 781
```

//进程 ID,路由器 ID,网络类型,接口 Cost 值。

```
Transmit Delay is 1 sec, State POINT－TO－POINT, Priority 0
```

//接口的延迟和状态。

```
No designated router on this network
No backup designated router on this network
Timer intervals configured, Hello 10, Dead 40, Wait 40, Retransmit 5
    Hello due in 00:00:07
```

```
    Index 2/2, flood queue length 0
    Next 0x0(0)/0x0(0)
    Last flood scan length is 1, maximum is 1
    Last flood scan time is 0 msec, maximum is 0 msec
Neighbor Count is 1 , Adjacent neighbor count is 1
```

//邻居的个数以及已建立邻接关系的邻居的个数。

```
  Adjacent with neighbor 192.168.0.1
```

//已经建立邻接关系的邻居路由器 ID。

```
  Suppress hello for 0 neighbor(s)
```

//没有进行 Hello 抑制。

```
Router1#show ip ospf neighbor

Neighbor ID     Pri    State      Dead Time    Address     Interface
192.168.0.1     0      FULL/ -    00:00:30     16.16.1.1   Serial2/0
192.168.2.1     0      FULL/ -    00:00:37     16.16.2.2   Serial3
```

//输出表明路由器 Router1 有两个邻居,它们的路由器 ID 分别为 1.1.1.1 和 3.3.3.3,其他参数解释如下。
//① Pri:邻居路由器接口的优先级。
//② State:当前邻居路由器接口的状态。
//③ Dead Time:清除邻居关系前等待的最长时间。
//④ Address:邻居接口的地址。
//⑤ Interface:自己和邻居路由器相连接口。
//⑥ "—":表示点到点的链路上 OSPF 不进行 DR 选举。

【提示】 OSPF 邻居关系不能建立的常见原因:

① hello 间隔和 dead 间隔不同。同一链路上的 hello 包间隔和 dead 间隔必须相同才能建立邻接关系。

② 区域号码不一致。

③ 特殊区域(如 stub,nssa 等)区域类型不匹配。

④ 认证类型或密码不一致。

⑤ 路由器 ID 相同。

⑥ Hello 包被 ACL deny。

⑦ 链路上的 MTU 不匹配。

⑧ 接口下 OSPF 网络类型不匹配。

```
Router1#show ip ospf database
         OSPF Router with ID (192.168.1.1) (Process ID 1)

                  Router Link States (Area 0)

Link ID          ADV Router       Age     Seq#         Checksum Link count
192.168.1.1      192.168.1.1      578     0x80000006   0x00feff   5
192.168.0.1      192.168.0.1      674     0x80000004   0x00feff   3
192.168.3.1      192.168.3.1      513     0x80000004   0x00feff   3
192.168.2.1      192.168.2.1      513     0x80000006   0x00feff   5
```

//以上输出是 Router1 的区域 0 的拓扑结构数据库的信息,标题行的解释如下。
//① Link ID:是指 Link State ID,代表整个路由器,而不是某个链路。
//② ADV Router:是指通告链路状态信息的路由器 ID。
//③ Age:老化时间。
//④ Seq#:序列号。
//⑤ Checksum:校验和。
//⑥ Link count:通告路由器在本区域内的链路数目。

6.10 实验 9 广播多路访问链路上的 OSPF

6.10.1 实验拓扑

根据提供路由器,配置广播多路访问链路上的 OSPF,并验证,见图 6.9。

设备名	IP 地址	端口
Router0	192.168.0.1	F0/0
Router0	16.16.1.1	F1/0
Router1	192.168.1.1	F0/0
Router1	16.16.1.2	F1/0
Router2	192.168.2.1	F0/0
Router2	16.16.1.3	F1/0
Router3	192.168.3.1	F0/0
Router3	16.16.1.4	F1/0
PC0	192.168.0.2	F0
PC1	192.168.1.2	F0
PC2	192.168.2.2	F0
PC3	192.168.3.2	F0

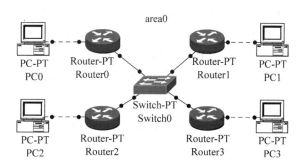

图 6.9 广播多路 OSPF

6.10.2 实验内容

(1) 在路由器上启动 OSPF 路由进程。
(2) 启用参与路由协议的接口,并且通告网络及所在的区域。
(3) 修改参考带宽。
(4) DR 选举的控制。

(5) 广播多路访问链路上的 OSPF 的特征。

6.10.3 实验步骤

(1) OSPF 路由配置。

```
Router0(config)# router ospf 1
Router0(config-router)# router-id 192.168.0.1
Router0(config-router)# net 192.168.0.0 255.255.255.0 area 0
Router0(config-router)# net 16.16.1.0 255.255.255.0 area 0
Router0(config-router)# auto-cost reference-bandwidth 1000
```

```
Router1(config)# router ospf 1
Router1(config-router)# router-id 192.168.1.1
Router1(config-router)# net 192.168.1.0 255.255.255.0 area 0
Router1(config-router)# net 16.16.1.0 255.255.255.0 area 0
Router1(config-router)# auto-cost reference-bandwidth 1000
```

```
Router2(config)# route ospf 1
Router2(config-router)# router-id 192.168.2.1
Router2(config-router)# net 192.168.2.0 255.255.255.0 area 0
Router2(config-router)# net 16.16.1.0 255.255.255.0 area 0
Router2(config-router)# auto-cost reference-bandwidth 1000
```

```
Router3(config)# router ospf 1
Router3(config-router)# router-id 192.168.3.1
Router3(config-router)# net 192.168.3.0 255.255.255.0 area 0
Router3(config-router)# net 16.16.1.0 255.255.255.0 area 0
Router3(config-router)# auto-cost reference-bandwidth 1000
```

【提示】 "auto-cost reference-bandwidth"命令是修改参考带宽的,因为本实验中的以太口的带宽为千兆,如果采用默认的百兆参考带宽,计算出来的 cost 是 0.1,这显然是不合理的。修改参考带宽要在所有的 OSPF 路由器上配置,目的是确保参考标准是相同的。另外,当执行命令"auto-cost reference-bandwidth"的时候,系统也会提示如下信息:

```
% OSPF: Reference bandwidth is changed.
Please ensure reference bandwidth is consistent across all routers.
```

(2) 验证调试。

```
Router0# show ip ospf neighbor

Neighbor ID     Pri   State          Dead Time    Address      Interface
192.168.1.1     1     FULL/BDR       00:00:32     16.16.1.2    FastEthernet1/0
192.168.2.1     1     FULL/DROTHER   00:00:33     16.16.1.3    FastEthernet1/0
192.168.3.1     1     FULL/DROTHER   00:00:32     16.16.1.4    FastEthernet1/0
```
//输出表明在该广播多路访问网络中,Router0 是 DR,Router1 是 BDR,Router2 和 Router3 为 DROTHER。

【提示】

① 为了避免路由器之间建立完全邻接关系而引起的大量开销，OSPF 要求在多路访问的网络中选举一个 DR，每个路由器都与之建立邻接关系。选举 DR 的同时也选举出一个 BDR，在 DR 失效的时候，BDR 担负起 DR 的职责，而且所有其他路由器只与 DR 和 BDR 建立邻接关系。

② DR 和 BDR 有它们自己的组播地址 224.0.0.6。

③ DR 和 BDR 的选举是以各个网络为基础的，也就是说，DR 和 BDR 选举是一个路由器的接口特性，而不是整个路由器的特性。

④ DR 选举的原则：

- 首要因素是时间，最先启动的路由器被选举成 DR。
- 如果同时启动，或者重新选举，则看接口优先级（范围为 0~255），优先级最高的被选举成 DR，默认情况下，多路访问网络的接口优先级为 1，点到点网络接口优先级为 0，修改接口优先级的命令是"ip ospf priority"，如果接口的优先级被设置为 0，那么该接口将不参与 DR 选举。
- 如果前两者相同，最后看路由器 ID，路由器 ID 最高的被选举成 DR。

⑤ DR 选举是非抢占的，除非人为地重新选举。重新选举 DR 的方法有两种，一是路由器重新启动，二是执行"clear ip ospf process"命令。

```
Router0#show ip ospf interface
FastEthernet0/0 is up, line protocol is up
  Internet address is 192.168.0.1/24, Area 0
  Process ID 1, Router ID 192.168.0.1, Network Type BROADCAST, Cost: 1
  Transmit Delay is 1 sec, State DR, Priority 1
```

// 自己 state 是 DR。

```
Designated Router (ID) 192.168.0.1, Interface address 192.168.0.1
```

// DR 的路由器 ID 以及接口地址。

```
  No backup designated router on this network
  Timer intervals configured, Hello 10, Dead 40, Wait 40, Retransmit 5
    Hello due in 00:00:03
  Index 1/1, flood queue length 0
  Next 0x0(0)/0x0(0)
  Last flood scan length is 1, maximum is 1
  Last flood scan time is 0 msec, maximum is 0 msec
  Neighbor Count is 0, Adjacent neighbor count is 0
  Suppress hello for 0 neighbor(s)
FastEthernet1/0 is up, line protocol is up
  Internet address is 16.16.1.1/24, Area 0
  Process ID 1, Router ID 192.168.0.1, Network Type BROADCAST, Cost: 1
  Transmit Delay is 1 sec, State DR, Priority 1
  Designated Router (ID) 192.168.0.1, Interface address 16.16.1.1
  Backup Designated Router (ID) 192.168.1.1, Interface address 16.16.1.2
  Timer intervals configured, Hello 10, Dead 40, Wait 40, Retransmit 5
```

```
        Hello due in 00:00:06
    Index 2/2, flood queue length 0
    Next 0x0(0)/0x0(0)
    Last flood scan length is 1, maximum is 1
    Last flood scan time is 0 msec, maximum is 0 msec
  Neighbor Count is 3, Adjacent neighbor count is 3
```

// Router0 是 DR,有三个邻居,并且全部形成邻接关系。

```
        Adjacent with neighbor 192.168.1.1 (Backup Designated Router)
        Adjacent with neighbor 192.168.2.1
        Adjacent with neighbor 192.168.3.1
  Suppress hello for 0 neighbor(s)
```

6.11 实验10 基于区域的 OSPF 简单口令认证

6.11.1 实验拓扑

根据提供路由器,配置基于区域的 OSPF 简单口令认证,见图 6.10。

设备名	IP 地址	端口
Router0	192.168.0.1	F0/0
Router0	16.16.1.1	S2/0
Router1	192.168.1.1	F0/0
Router1	16.16.1.2	S2/0
PC0	192.168.0.2	F0
PC1	192.168.1.2	F0

6.11.2 实验内容

(1) 掌握 OSPF 认证的类型和意义。
(2) 掌握基于区域的 OSPF 简单口令认证的配置。
(3) 验证调试。

6.11.3 实验步骤

(1) 配置路由器。

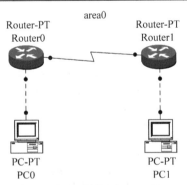

图 6.10 区域 OSPF 口令认证

```
Router0(config)# router ospf 1
Router0(config-router)# router-id 192.168.0.1
Router0(config-router)# net 16.16.1.0 255.255.255.0 area 0
Router0(config-router)# net 192.168.0.0 255.255.255.0 area 0
Router0(config-router)# area 0 authentication
```

//区域 0 启用简单口令认证。

```
Router0(config-router)# int s2/0
Router0(config-if)# ip ospf authentication-key cisco
```

//配置认证密码。

```
Router1(config-if)# router ospf 1
Router1(config-router)# router-id 192.168.1.1
Router1(config-router)# net 16.16.1.0 255.255.255.0 area 0
Router1(config-router)# net 192.168.1.0 255.255.255.0 area 0
Router1(config-router)# area 0 authentication
Router1(config-router)# int s2/0
Router1(config-if)# ip ospf authentication-key cisco
```

（2）验证调试。

```
Router0# show ip ospf interface s2/0
Serial2/0 is up, line protocol is up
  Internet address is 16.16.1.1/24, Area 0
  Process ID 1, Router ID 192.168.0.1, Network Type POINT-TO-POINT, Cost: 781
  Transmit Delay is 1 sec, State POINT-TO-POINT, Priority 0
  No designated router on this network
  No backup designated router on this network
  Timer intervals configured, Hello 10, Dead 40, Wait 40, Retransmit 5
    Hello due in 00:00:00
  Index 1/1, flood queue length 0
  Next 0x0(0)/0x0(0)
  Last flood scan length is 1, maximum is 1
  Last flood scan time is 0 msec, maximum is 0 msec
  Neighbor Count is 1 , Adjacent neighbor count is 1
    Adjacent with neighbor 192.168.1.1
  Suppress hello for 0 neighbor(s)
Simple password authentication enabled
```

//输出信息表明该接口启用了简单口令认证。

```
Router0# show ip ospf
Routing Process "ospf 1" with ID 192.168.0.1
Supports only single TOS(TOS0) routes
Supports opaque LSA
SPF schedule delay 5 secs, Hold time between two SPFs 10 secs
Minimum LSA interval 5 secs. Minimum LSA arrival 1 secs
Number of external LSA 0. Checksum Sum 0x000000
Number of opaque AS LSA 0. Checksum Sum 0x000000
Number of DCbitless external and opaque AS LSA 0
Number of DoNotAge external and opaque AS LSA 0
Number of areas in this router is 1. 1 normal 0 stub 0 nssa
External flood list length 0
    Area BACKBONE(0)
        Number of interfaces in this area is 2
        Area has simple password authentication
        SPF algorithm executed 3 times
```

```
Area ranges are
Number of LSA 2. Checksum Sum 0x0097b3
Number of opaque link LSA 0. Checksum Sum 0x000000
Number of DCbitless LSA 0
Number of indication LSA 0
Number of DoNotAge LSA 0
Flood list length 0
```

//输出表明区域0采用简单口令认证。

① 如果 Router0 区域0没有启动认证,而 Router1 区域0启动简单口令认证,则 Router1 上出现下面的信息:

```
OSPF: Rcv pkt from 16.16.1.1, Serial2/0 : Mismatch Authentication type. Input packet specified type 0, we use type 1
```

② 如果 Router0 和 Router1 的区域0都启动简单口令认证,但是 Router1 的接口下没有配置密码或密码错误,则 Router1 上出现下面的信息:

```
OSPF: Rcv pkt from 16.16.1.1, Serial2/0 : Mismatch Authentication Key - Clear Text
```

6.12 实验11 基于区域的 OSPF MD5 认证

6.12.1 实验拓扑

根据提供路由器,配置基于区域的 OSPF MD5 口令认证,见图 6.11。

设备名	IP 地址	端口
Router0	192.168.0.1	F0/0
Router0	16.16.1.1	S2/0
Router1	192.168.1.1	F0/0
Router1	16.16.1.2	S2/0
PC0	192.168.0.2	F0
PC1	192.168.1.2	F0

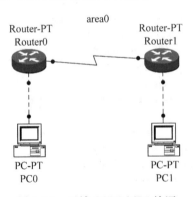

图 6.11 区域 OSPF MD5 认证

6.12.2 实验内容

(1) 掌握 OSPF 认证的类型和意义。
(2) 掌握基于区域的 OSPF MD5 认证的配置。
(3) 验证调试。

6.12.3 实验步骤

(1) 路由器配置。

```
Router0(config)#router ospf 1
Router0(config-router)#area 0 authentication message-digest
```

//区域 0 启用 MD5 认证。

```
Router0(config-router)#int s2/0
Router0(config-if)#ip ospf message-digest-key 1 md5 cisco
```

//配置认证 key ID 及密匙。

```
Router1(config)#router ospf 1
Router1(config-router)#area 0 authentication message-digest
Router1(config-router)#int s2/0
Router1(config-if)#ip ospf message-digest-key 1 md5 cisco
```

(2) 验证调试。

```
Router0#sh ip ospf interface s2/0
Serial2/0 is up, line protocol is up
  Internet address is 16.16.1.1/24, Area 0
  Process ID 1, Router ID 192.168.0.1, Network Type POINT-TO-POINT, Cost: 781
  Transmit Delay is 1 sec, State POINT-TO-POINT, Priority 0
  No designated router on this network
  No backup designated router on this network
  Timer intervals configured, Hello 10, Dead 40, Wait 40, Retransmit 5
    Hello due in 00:00:09
  Index 1/1, flood queue length 0
  Next 0x0(0)/0x0(0)
  Last flood scan length is 1, maximum is 1
  Last flood scan time is 0 msec, maximum is 0 msec
  Neighbor Count is 1 , Adjacent neighbor count is 1
    Adjacent with neighbor 192.168.1.1
  Suppress hello for 0 neighbor(s)
  Message digest authentication enabled
  Youngest key id is 1
```

//输出信息表明该接口启用了 MD5 认证,而且密钥 ID 为 1。

```
Router0#show ip ospf
Routing Process "ospf 1" with ID 192.168.0.1
Supports only single TOS(TOS0) routes
```

```
Supports opaque LSA
SPF schedule delay 5 secs, Hold time between two SPFs 10 secs
Minimum LSA interval 5 secs. Minimum LSA arrival 1 secs
Number of external LSA 0. Checksum Sum 0x000000
Number of opaque AS LSA 0. Checksum Sum 0x000000
Number of DCbitless external and opaque AS LSA 0
Number of DoNotAge external and opaque AS LSA 0
Number of areas in this router is 1. 1 normal 0 stub 0 nssa
External flood list length 0
    Area BACKBONE(0)
        Number of interfaces in this area is 2
        Area has message digest authentication
        SPF algorithm executed 5 times
        Area ranges are
        Number of LSA 2. Checksum Sum 0x008fb7
        Number of opaque link LSA 0. Checksum Sum 0x000000
        Number of DCbitless LSA 0
        Number of indication LSA 0
        Number of DoNotAge LSA 0
        Flood list length 0
```

//输出表明区域 0 采用 MD5 认证。

① 如果 Router0 区域 0 启动 MD5 认证，而 Router1 区域 0 启动简单口令认证，则 Router1 上出现下面的信息：

```
OSPF: Rcv pkt from 16.16.1.1, Serial2/0 : Mismatch Authentication type. Input packet specified type 2, we use type 1
```

② 如果 Router0 和 Router1 的区域 0 都启动 MD5 认证，但是 Router1 的接口下没有配置 key ID 和密码或密码错误，则 Router1 上出现下面的信息：

```
OSPF: Rcv pkt from 16.16.1.1, Serial2/0 : Mismatch Authentication Key - No message digest key 1 on interface
```

第7章　交换机基本配置

7.1　基础知识

交换机是第2层网络设备,主要是作为工作站、服务器、路由器、集线器和其他交换机的集中点。交换机是一台多端口的网桥,是当前采用星状拓扑结构的以太局域网的标准技术。交换机为所连接的两台的联网设备提供一条独享的点到点的虚线路,因此避免了冲突。交换机可以工作在全双工模式之下,这意味着可以同时发送和接收数据。

7.1.1　交换机工作原理

交换机根据收到数据帧中的源MAC地址建立该地址同交换机端口的映射,并将其写入MAC地址表中。交换机将数据帧中的目的MAC地址同已建立的MAC地址表进行比较,以决定由哪个端口进行转发。如果数据帧中的目的MAC地址不在MAC地址表中,则向所有端口转发。这一过程称为泛洪(flood)。广播帧和组播帧向所有的端口转发。

7.1.2　交换机功能

以太网交换机了解每一端口相连设备的MAC地址,并将地址同相应的端口映射起来存放在交换机缓存中的MAC地址表中。当一个数据帧的目的地址在MAC地址表中有映射时,它被转发到连接目的节点的端口而不是所有端口(如该数据帧为广播/组播帧则转发至所有端口)。当交换机包括一个冗余回路时,以太网交换机通过生成树协议避免回路的产生,同时允许存在后备路径。

7.1.3　交换机工作特性

交换机的每一个端口所连接的网段都是一个独立的冲突域。交换机所连接的设备仍然在同一个广播域内,也就是说,交换机不隔绝广播(唯一的例外是在配有VLAN的环境中)。交换机依据帧头的信息进行转发,因此说交换机是工作在数据链路层的网络设备(此处所述交换机仅指传统的二层交换设备)。

7.1.4　交换机分类

依照交换机处理帧时不同的操作模式,主要可分为两类。
(1) 存储转发:交换机在转发之前必须接收整个帧,并进行错误校检,如无错误再将这一帧发往目的地址。帧通过交换机的转发时延随帧长度的不同而变化。

(2) 直通式：交换机只要检查到帧头中所包含的目的地址就立即转发该帧，而无须等待帧全部被接收，也不进行错误校验。由于以太网帧头的长度总是固定的，因此帧通过交换机的转发时延也保持不变。

7.2　实验1　控制台方式访问交换机

7.2.1　实验拓扑

一台 PC 通过 RS-232 串口使用控制台线连接一台交换机的 Console 端口，见图 7.1。

7.2.2　实验内容

(1) 通过控制台线缆实现 PC 连接交换机。
(2) 通过控制台程序实现 PC 访问交换机。

图 7.1　控制台方式连接交换机

7.2.3　实验步骤

(1) 选择【开始】→【所有程序】→【附件】→【通讯】→【超级终端】菜单项，启动超级终端程序，见图 7.2。

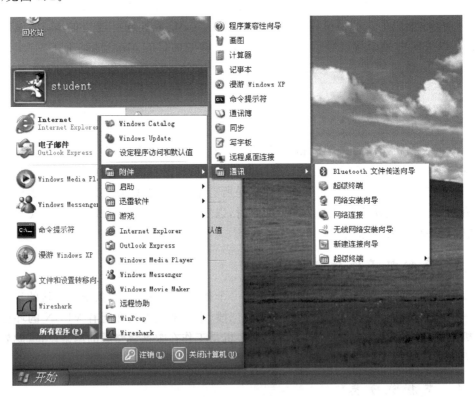

图 7.2　启动超级终端

(2) 输入新建连接名称为 cisco，选择任一图标，单击【确定】按钮，见图 7.3。
(3) 根据在连接时使用列表选择 COM1，单击【确定】按钮，见图 7.4。

图 7.3 输入连接名称

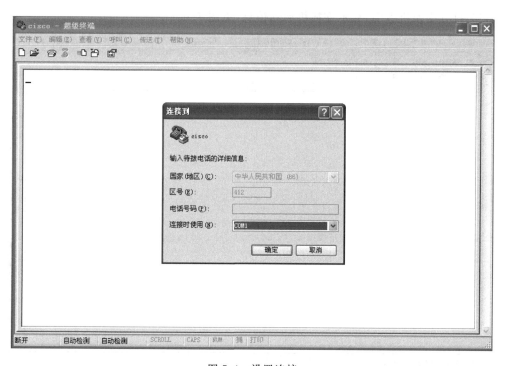

图 7.4 设置连接

（4）单击【还原为默认值】按钮，还原 COM1 端口设置为默认属性值，单击【确定】按钮，见图 7.5。

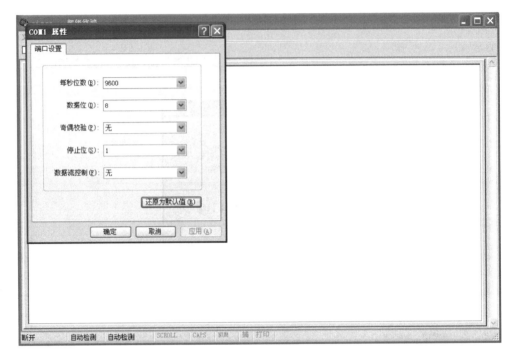

图 7.5 还原为默认值

(5) 打开交换机电源,显示交换机启动信息,见图 7.6。

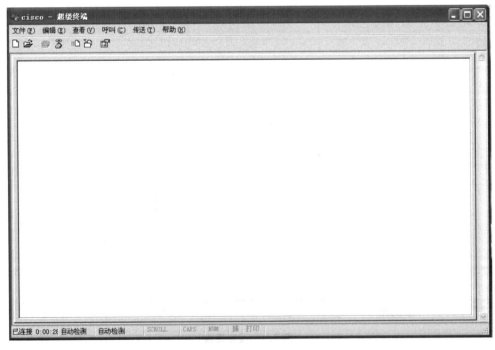

图 7.6 显示交换机启动信息

7.3 实验 2 交换机基本配置

7.3.1 实验拓扑

一台 PC 通过 RS-232 串口使用控制台线连接一台交换机的 Console 端口，见图 7.7。

7.3.2 实验内容

（1）查看 IOS 版本信息、默认配置信息、端口属性信息和 VLAN 属性信息。

图 7.7 控制台方式连接交换机

（2）配置交换机主机名、口令和管理 IP。
（3）保存配置。
（4）重新启动交换机，验证配置。

7.3.3 实验步骤

（1）进入特权模式。

```
    Press RETURN to get started!

Switch> enable
Switch#
```

（2）查看 IOS 版本信息。

```
    Switch# show version
Cisco Internetwork Operating System Software
IOS (tm) C2950 Software (C2950 - I6Q4L2 - M), Version 12.1(22)EA4, RELEASE SOFTWARE(fc1)
Copyright (c) 1986 - 2005 by cisco Systems, Inc.
Compiled Wed 18 - May - 05 22:31 by jharirba
Image text - base: 0x80010000, data - base: 0x80562000

ROM: Bootstrap program is is C2950 boot loader

Switch uptime is 2 minutes, 45 seconds
System returned to ROM by power - on

Cisco WS - C2950 - 24 (RC32300) processor (revision C0) with 21039K bytes of memory.
Processor board ID FHK0610Z0WC
Last reset from system - reset
Running Standard Image
24 FastEthernet/IEEE 802.3 interface(s)

63488K bytes of flash - simulated non - volatile configuration memory.
Base ethernet MAC Address: 0050.0FED.46D7
Motherboard assembly number: 73 - 5781 - 09
Power supply part number: 34 - 0965 - 01
```

```
Motherboard serial number: FOC061004SZ
Power supply serial number: DAB0609127D
Model revision number: C0
Motherboard revision number: A0
-- More --
```

【提示】 按【回车】键,再显示一行信息,按【空格】键,再显示一页信息,按 Esc 键,结束显示信息。

(3) 查看默认配置信息。

```
Switch# show running-config
Building configuration...

Current configuration : 971 bytes
!
version 12.1
no service timestamps log datetime msec
no service timestamps debug datetime msec
no service password-encryption
!
!
hostname Switch
!
!
spanning-tree mode pvst
!
interface FastEthernet0/1
!
interface FastEthernet0/2
!
interface FastEthernet0/3
!
interface FastEthernet0/4
!
-- More --
```

【提示】 按【回车】键,再显示一行信息,按【空格】键,再显示一页信息,按 Esc 键,结束显示信息。

(4) 查看端口属性信息。

```
Switch# show interfaces
FastEthernet0/1 is down, line protocol is down (disabled)
  Hardware is Lance, address is 00d0.588d.3601 (bia 00d0.588d.3601)
  BW 100000 Kbit, DLY 1000 usec,
     reliability 255/255, txload 1/255, rxload 1/255
  Encapsulation ARPA, loopback not set
  Keepalive set (10 sec)
  Half-duplex, 100Mb/s
  input flow-control is off, output flow-control is off
```

```
ARP type: ARPA, ARP Timeout 04:00:00
Last input 00:00:08, output 00:00:05, output hang never
Last clearing of "show interface" counters never
Input queue: 0/75/0/0 (size/max/drops/flushes); Total output drops: 0
Queueing strategy: fifo
Output queue :0/40 (size/max)
5 minute input rate 0 bits/sec, 0 packets/sec
5 minute output rate 0 bits/sec, 0 packets/sec
    956 packets input, 193351 bytes, 0 no buffer
    Received 956 broadcasts, 0 runts, 0 giants, 0 throttles
    0 input errors, 0 CRC, 0 frame, 0 overrun, 0 ignored, 0 abort
    0 watchdog, 0 multicast, 0 pause input
    0 input packets with dribble condition detected
    2357 packets output, 263570 bytes, 0 underruns
-- More --
```

【提示】 按【回车】键,再显示一行信息,按【空格】键,再显示一页信息,按 Esc 键,结束显示信息。

(5) 查看 VLAN 属性信息。

```
Switch# show vlan

VLAN Name                             Status      Ports
---- -------------------------------- ----------- -------------------------------
1    default                          active      Fa0/1, Fa0/2, Fa0/3, Fa0/4
                                                  Fa0/5, Fa0/6, Fa0/7, Fa0/8
                                                  Fa0/9, Fa0/10, Fa0/11, Fa0/12
                                                  Fa0/13, Fa0/14, Fa0/15, Fa0/16
                                                  Fa0/17, Fa0/18, Fa0/19, Fa0/20
                                                  Fa0/21, Fa0/22, Fa0/23, Fa0/24
1002 fddi-default                     act/unsup
1003 token-ring-default               act/unsup
1004 fddinet-default                  act/unsup
1005 trnet-default                    act/unsup

VLAN Type  SAID    MTU   Parent RingNo BridgeNo Stp  BrdgMode Trans1 Trans2
---- ----- ------- ----- ------ ------ -------- ---- -------- ------ ------
1    enet  100001  1500  -      -      -        -    -        0      0
1002 fddi  101002  1500  -      -      -        -    -        0      0
1003 tr    101003  1500  -      -      -        -    -        0      0
1004 fdnet 101004  1500  -      -      -        ieee -        0      0
1005 trnet 101005  1500  -      -      -        ibm  -        0      0

-- More --
```

【提示】 按【回车】键,再显示一行信息,按【空格】键,再显示一页信息,按 Esc 键,结束显示信息。

(6) 进入全局配置模式。

```
Switch# configure terminal
Enter configuration commands, one per line. End with CNTL/Z.
Switch(config)#
```

(7) 配置主机名为 S1。

```
Switch(config)# hostname S1
S1(config)#
```

【提示】 主机名从 Switch 变为 S1。

(8) 配置特权模式密码为 cisco。

```
S1(config)# enable secret cisco
S1(config)#
```

【提示】 在不支持密文密码的机器上,也可以配置明文密码,使用 enable password cisco 命令,如果同时配置了密文密码和明文密码,密文密码有效。

(9) 配置管理 IP。

```
    S1(config)# interface vlan 1
S1(config-if)# ip address 192.168.1.1 255.255.255.0
S1(config-if)# no shutdown
S1(config-if)#
```

【提示】 管理 VLAN 端口和物理端口一样,需要启用配置才能生效。

(10) 返回特权模式。

```
S1(config-if)# end
S1#
%SYS-5-CONFIG_I: Configured from console by console
```

【提示】 可以直接使用 end 命令回到特权模式,也可以使用 exit 命令逐层返回上一模式,直至特权模式。

(11) 保存配置。

```
S1# copy running-config startup-config
Destination filename [startup-config]?
Building configuration...
[OK]
S1#
```

【提示】 只有保存运行配置,才能当机器重新启动后,运行配置中配置还能继续有效。

(12) 重新启动交换机。

```
S1#reload
Proceed with reload? [confirm]
```

(13) 验证主机名。

```
    Press RETURN to get started!

%LINK-5-CHANGED: Interface Vlan1, changed state to up

%SYS-5-CONFIG_I: Configured from console by console

S1>
```

(14) 验证特权模式密码。

进入特权模式提示输入密码时,输入 cisco。

```
S1>enable
Password:
S1#
```

【提示】 输入特权模式密码 cisco,命令行并不回显 cisco 密码字符。

(15) 验证管理 IP。

配置一台 PC 的 IP 地址为 192.168.1.2,子网掩码为 255.255.255.0,用网线连接交换机任意快速以太网网口,使用 ping 命令测试到交换机连通性。

```
PC>ping 192.168.1.1

Pinging 192.168.1.1 with 32 bytes of data:

Reply from 192.168.1.1: bytes=32 time=5ms TTL=255
Reply from 192.168.1.1: bytes=32 time=5ms TTL=255
Reply from 192.168.1.1: bytes=32 time=3ms TTL=255
Reply from 192.168.1.1: bytes=32 time=5ms TTL=255

Ping statistics for 192.168.1.1:
    Packets: Sent = 4, Received = 4, Lost = 0 (0% loss),
Approximate round trip times in milli-seconds:
    Minimum = 3ms, Maximum = 5ms, Average = 4ms
```

7.4 实验3 Telnet 方式访问交换机

7.4.1 实验拓扑

PC 通过 RS-232 串口使用控制台线连接交换机 Switch1 的 Console 端口,PC 和

Switch2 交换机通过快速以太网网口连接 Switch1 交换机的快速以太网网口，见图 7.8。

设备名	IP 地址
Switch1	192.168.1.1
Switch2	192.168.1.2
PC	192.168.1.3

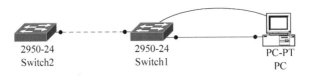

图 7.8　Telnet 方式连接交换机

7.4.2　实验内容

(1) 控制台方式配置 Switch1 支持 Telnet 访问。

(2) PC 和 Switch2 分别 Telnet 访问 Switch1。

7.4.3　实验步骤

(1) 通过控制台方式配置虚拟终端线路。

```
Switch1 > enable
Switch1 # configure terminal
Enter configuration commands, one per line. End with CNTL/Z.
Switch1(config) # line vty 0 15
Switch1(config - line) # password cisco
Switch1(config - line) # login
Switch1(config - line) #
```

【提示】　由于不同型号的交换机虚拟终端线路范围不同，为了保持所有虚拟终端线路配置相同，建议先通过 line vty 0 ? 命令查询最大虚拟终端线路编号，然后对所有虚拟终端线路一起进行配置。

(2) PC 和 Switch2 分别 Telnet 访问 Switch1。

```
PC:
PC> telnet 192.168.1.1
Trying 192.168.1.1 ...Open


User Access Verification

Password:
Switch1 >


Switch2:
Switch2 # telnet 192.168.1.1
```

```
Trying 192.168.1.1 ...Open


User Access Verification

Password:
Switch1>
```

【提示】 输入 Telnet 密码 cisco,命令行并不回显 cisco 密码字符,主机名标识符变为 Switch1,说明当前已经登录到 Switch1,以后输入的命令将在 Switch1 上执行。

(3) 查看 Switch1 用户连接信息。

```
Switch1 # show users
    Line      User     Host(s)        Idle       Location
*  0 con 0              idle          00:00:00
   2 vty 0              idle          00:00:12   192.168.1.3
   3 vty 1              idle          00:00:04   192.168.1.2

  Interface  User     Mode           Idle       Peer Address
```

【提示】 一个用户从控制台线路连接,会话编号为 0,标记为当前会话;一个用户从 IP 地址为 192.168.1.3 的终端(PC)连接虚拟终端线路 0,会话编号为 2;一个用户从 IP 地址为 192.168.1.2 的终端(Switch2)连接虚拟终端线路 1,会话编号为 3。

(4) 挂起 Switch2 当前会话信息,同时按 Shift+Ctrl+6 键后,抬起,按 X 键。

```
Switch1>
Switch2#
```

【提示】 主机名标识符从 Switch1 变为 Switch2,说明连接到 Switch1 的会话暂时被挂起,当前会话返回到 Switch2,以后输入的命令将在 Switch2 上执行。

(5) 查看 Switch2 会话信息。

```
Switch2 # show sessions
Conn Host           Address        Byte    Idle Conn Name
*  1 192.168.1.1    192.168.1.1    0       3    192.168.1.1
```

【提示】 Switch2 发起到 IP 地址为 192.168.1.1(Switch1)的会话,会话编号为 1,标记为当前会话。

(6) 恢复 Switch2 指定会话信息。

```
Switch2 # resume 1
[Resuming connection 1 to 192.168.1.1 ... ]

Switch1>
```

【提示】 主机名提示符从 Switch2 变为 Switch1,说明连接到 Switch1 的会话被恢复,当前会话返回到 Switch1，以后输入的命令将在 Switch1 上执行。

(7) 挂起 Switch2 当前会话，关闭指定会话，再次查看 Switch2 会话信息。

```
Switch1 >
Switch2 # disconnect 1
Closing connection to 192.168.1.1 [confirm]
```

```
Switch # show sessions
 % No connections open
```

【提示】 Switch2 挂起到 Switch1 的当前会话，当前会话返回到 Switch2，执行命令主动中断自己发起的到 Switch1 的会话，会话编号为 1。

(8) Switch1 清除指定线路信息，再次查看用户连接信息。

```
Switch1 # clear line 2
[confirm]
Switch1 # show users
  Line       User      Host(s)       Idle         Location
*  0 con 0              idle          00:00:00

  Interface  User      Mode          Idle         Peer Address
```

【提示】 Switch1 执行命令主动中断别人发起的到 Switch1 会话，会话编号为 2（虚拟终端线路编号为 0），查看用户连接信息，只剩下控制台线路连接信息。

7.5 实验 4 MAC 地址管理

7.5.1 实验拓扑

PC1 连接 Switch 交换机的快速以太网网口 Fa0/2，PC2 和 PC3 通过集线器连接 Switch 交换机的快速以太网网口 Fa0/1，见图 7.9。

设备名	IP 地址	MAC 地址
PC1	192.168.1.1	0009.7C70.C10E
PC2	192.168.1.2	0050.0F81.2830
PC3	192.168.1.3	00E0.A334.A98D
PC4	192.168.1.4	0009.7C03.D142
PC5	192.168.1.5	0001.9685.9D16

7.5.2 实验内容

(1) 查看 MAC 地址表。
(2) 配置静态 MAC 地址。
(3) 配置端口安全。

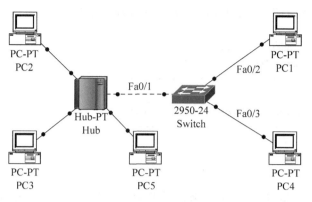

图 7.9 MAC 地址管理

7.5.3 实验步骤

(1) 查看初始 MAC 地址表。

```
Switch>
Switch> enable
Switch# show mac-address-table
          Mac Address Table
-------------------------------------------

Vlan    Mac Address      Type        Ports
----    -----------      --------    -----


Switch#
```

【提示】 初始时,没有发生任何经过 Switch 的通信,所以 Switch 没有学到任何 MAC 地址信息。

(2) PC2 和 PC3 分别 ping PC1,再次查看 MAC 地址表。

```
Switch# show mac-address-table
          Mac Address Table
-------------------------------------------

Vlan    Mac Address      Type        Ports
----    -----------      --------    -----
 1      0009.7c70.c10e   DYNAMIC     Fa0/2
 1      0050.0f81.2830   DYNAMIC     Fa0/1
 1      00e0.a334.a98d   DYNAMIC     Fa0/1
Switch#
```

【提示】 由于 PC2 和 PC3 的 ping 包都经过 Switch 的 Fa0/1 端口,所以 Switch 在 Fa0/1 端口分别动态学到 PC2 和 PC3 的 MAC 地址。由于 PC1 的 ping 响应包经过 Switch 的 Fa0/2 端口,所以 Switch 在 Fa0/2 端口动态学到 PC1 的 MAC 地址。

(3) 不进行任何通信,一段时间后再次查看 MAC 地址表。

```
Switch# show mac-address-table
        Mac Address Table
-------------------------------------------

Vlan    Mac Address     Type        Ports
----    -----------     --------    -----

Switch#
```

【提示】 由于 MAC 地址表项是动态学习的,会因为老化时间超时被自动从 MAC 地址表清除。

(4) 添加 PC4 的 MAC 地址到 MAC 地址表。

```
Switch# config terminal
Enter configuration commands, one per line. End with CNTL/Z.
Switch(config)# mac-address-table static 0009.7C03.D142 vlan 1 interface fa0/3
Switch(config)#
Switch(config)# end
Switch#
% SYS-5-CONFIG_I: Configured from console by console
```

(5) 再次查看 MAC 地址表。

```
Switch# show mac-address-table
        Mac Address Table
-------------------------------------------

Vlan    Mac Address     Type        Ports
----    -----------     --------    -----

  1     0009.7c03.d142  STATIC      Fa0/3
Switch#
```

【提示】 手动添加的 MAC 地址表项类型为静态,不会因为老化时间超时被从 MAC 地址表自动清除,只能手动清除。

(6) Fa0/1 端口启用端口安全。

```
Switch# config terminal
Enter configuration commands, one per line. End with CNTL/Z.
Switch(config)# interface fastEthernet 0/1
Switch(config-if)# switchport mode access
Switch(config-if)# switchport port-security
Switch(config-if)# end
Switch#
% SYS-5-CONFIG_I: Configured from console by console
```

【提示】 动态端口不能启动端口安全性,只有静态接入端口或者干道端口才可以。

(7) 清空 MAC 地址表,PC2 ping PC1,再次查看 MAC 地址表和关于 Fa0/1 的运行配置。

```
Switch# clear mac-address-table
Switch# show mac-address-table
          Mac Address Table
-------------------------------------------

Vlan     Mac Address       Type        Ports
----     -----------       --------    -----

  1      0009.7c70.c10e    DYNAMIC     Fa0/2
  1      0050.0f81.2830    STATIC      Fa0/1
Switch# show running-config
Building configuration...

interface FastEthernet0/1
switchport mode access
switchport port-security
!
```

【提示】 Fa0/1 端口启用端口安全,学到的 MAC 地址表项类型为静态;Fa0/2 端口未启用端口安全,为动态端口,学到的 MAC 地址表项类型为动态。Fa0/1 端口未启用 MAC 地址黏性学习,所以学到的 MAC 地址不会添加到 Fa0/1 端口相关的运行配置中,当交换机重新启动后,需要重新学习该 MAC 地址。

(8) Fa0/1 端口安全地址为黏性安全地址。

```
Switch# config terminal
Enter configuration commands, one per line. End with CNTL/Z.
Switch(config)# interface fastEthernet 0/1
Switch(config-if)# switchport port-security mac-address sticky
Switch(config-if)# end
Switch#
% SYS-5-CONFIG_I: Configured from console by console
```

(9) 清空 MAC 地址表,PC2 ping PC1,再次查看 MAC 地址表和关于 Fa0/1 的运行配置。

```
Switch# clear mac-address-table
Switch# show mac-address-table
          Mac Address Table
-------------------------------------------

Vlan     Mac Address       Type        Ports
----     -----------       --------    -----

  1      0009.7c70.c10e    DYNAMIC     Fa0/2
  1      0050.0f81.2830    STATIC      Fa0/1
Switch# show running-config
```

```
Building configuration...

interface FastEthernet0/1
switchport mode access
switchport port - security
switchport port - security mac - address sticky
switchport port - security mac - address sticky 0050.0F81.2830
```

【提示】 Fa0/1 端口启用 MAC 地址黏性学习,所以学到的 MAC 地址会被添加到 Fa0/1 端口相关的运行配置中,可通过保存运行配置,当交换机重新启动后,不用再重新学习该 MAC 地址。

(10) Fa0/1 端口安全地址最大数为 2。

```
Switch#config terminal
Enter configuration commands, one per line. End with CNTL/Z.
Switch(config)#interface fastEthernet 0/1
Switch(config - if)#switchport port - security maximum 2
```

【提示】 启用端口安全端口的默认安全地址最大数为 1。

(11) Fa0/1 端口安全违反处理为关闭端口。

```
Switch(config - if)#switchport port - security violation shutdown
Switch(config - if)#end
Switch#
%SYS - 5 - CONFIG_I: Configured from console by console
```

【提示】 端口安全违反处理包括 protect、shutdown 和 restrict。

protect:当端口上的安全 MAC 地址数量达到允许的极限时,带有未知源地址的包被丢弃,直到使用者删除足够多的安全 MAC 地址,或增加允许的最多地址数,但是端口不会被关闭,交换机不发送警告信息。

shutdown:当端口上的安全 MAC 地址数量达到允许的极限时,带有未知源地址的包使端口立即进入错误禁止状态,而且端口被关闭,交换机发送警告信息。

当端口处于错误禁止状态,使用者可用下列方法使它离开该状态:

① 使用全局配置模式命令 errdisable recovery cause psecure-violation。

② 使用端口模式命令 shutdown 和 no shutdown 关闭后再重新启用。

restrict:当在端口上的安全 MAC 地址数量达到允许的极限时,带有未知源地址的包被丢弃,直到使用者删除足够多的安全 MAC 地址,或增加允许的最多地址数,但是端口不会被关闭,交换机发送警告信息。

(12) 清空 MAC 地址表,PC2、PC3 和 PC5 分别 ping PC1。

```
PC2、PC3:
PC>ping 192.168.1.1

Pinging 192.168.1.1 with 32 bytes of data:
```

```
Reply from 192.168.1.1: bytes = 32 time = 33ms TTL = 128
Reply from 192.168.1.1: bytes = 32 time = 10ms TTL = 128
Reply from 192.168.1.1: bytes = 32 time = 14ms TTL = 128
Reply from 192.168.1.1: bytes = 32 time = 10ms TTL = 128

Ping statistics for 192.168.1.1:
    Packets: Sent = 4, Received = 4, Lost = 0 (0% loss),
Approximate round trip times in milli-seconds:
    Minimum = 10ms, Maximum = 33ms, Average = 16ms
```

```
      PC5:
PC > ping 192.168.1.1

Pinging 192.168.1.1 with 32 bytes of data:

Request timed out.
Request timed out.
Request timed out.
Request timed out.

Ping statistics for 192.168.1.1:
Packets: Sent = 4, Received = 0, Lost = 4 (100% loss),
    Switch:
% LINK - 5 - CHANGED: Interface FastEthernet0/1, changed state to administratively down

% LINEPROTO - 5 - UPDOWN: Line protocol on Interface FastEthernet0/1, changed state to down
```

【提示】 PC2 和 PC3 ping PC1 时,刚好达到 Fa0/1 端口安全地址最大数 2,没有触发端口安全违反处理,PC2 和 PC3 能够 ping 通 PC1。但是当 PC5 ping PC1 时,超过 Fa0/1 端口安全地址最大数 2,触发端口安全违反处理,Fa0/1 端口进入错误禁止状态,端口被关闭,PC5 不能够 ping 通 PC1。

(13) 查看端口安全信息。

```
Switch# show port - security
Secure Port MaxSecureAddr CurrentAddr SecurityViolation Security Action
              (Count)        (Count)      (Count)
-------------------------------------------------------------------------
    Fa0/1        2             2             1          Shutdown
-------------------------------------------------------------------------

Switch# show port - security address
            Secure Mac Address Table
-------------------------------------------------------------------------
Vlan    Mac Address        Type          Ports           Remaining Age
                                                         (mins)
----    -----------        ----          -----           -------------
1       0050.0F81.2830     SecureSticky  FastEthernet0/1      -
1       00E0.A334.A98D     SecureSticky  FastEthernet0/1      -
-------------------------------------------------------------------------
```

```
Total Addresses in System (excluding one mac per port)        : 1
Max Addresses limit in System (excluding one mac per port) : 1024

Switch# show port-security interface fastEthernet 0/1
Port Security                       : Enabled
Port Status                         : Secure-shutdown
Violation Mode                      : Shutdown
Aging Time                          : 0 mins
Aging Type                          : Absolute
SecureStatic Address Aging          : Disabled
Maximum MAC Addresses               : 2
Total MAC Addresses                 : 2
Configured MAC Addresses            : 0
Sticky MAC Addresses                : 2
Last Source Address:Vlan            : 0001.9685.9D16:1
Security Violation Count            : 1
```

【提示】 安全违法计数增1,从0变为1。

(14) Fa0/1端口关闭再打开,恢复正常工作状态。

```
Switch# configure terminal
Enter configuration commands, one per line. End with CNTL/Z.
Switch(config)# interface fastEthernet 0/1
Switch(config-if)# shutdown

% LINK-5-CHANGED: Interface FastEthernet0/1, changed state to administratively down
Switch(config-if)# no shutdown

% LINK-5-CHANGED: Interface FastEthernet0/1, changed state to up

% LINEPROTO-5-UPDOWN: Line protocol on Interface FastEthernet0/1, changed state to up
Switch(config-if)# end
Switch#
% SYS-5-CONFIG_I: Configured from console by console
```

【提示】 直接在Fa0/1端口执行no shutdown命令是不能离开端口错误禁止状态的,必须先执行shutdown命令关闭端口,然后再执行no shutdown命令才可以恢复端口正常状态。

第 8 章　虚拟局域网

8.1　基础知识

8.1.1　VLAN 定义

VLAN(Virtual Local Area Network)即虚拟局域网,是一种通过将局域网内的设备逻辑地而不是物理地划分成一个个网段从而实现虚拟工作组的技术,如图 8.1 所示。创建 VLAN 可以不受物理位置限制而根据用户需求进行网络分段,是对连接到第二层交换机端口的网络用户的逻辑分段,这样可以方便地按部门来分割网段,实现按权限访问,提高了安全性。

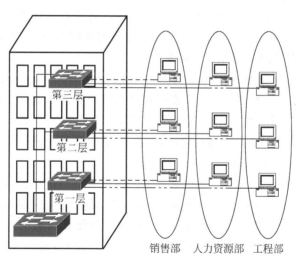

图 8.1　逻辑定义网络 VLAN

基于交换机的虚拟局域网能够为局域网解决冲突域、广播域、带宽问题。传统的共享介质的以太网和交换式的以太网中,对广播风暴的控制和网络安全只能在第三层的路由器上实现,而 VLAN 技术的出现可以在第二层上实现对广播域的划分。

划分 VLAN 后,由于广播域的缩小,网络中广播包消耗带宽所占的比例大大降低,网络的性能得到显著的提高。网络管理员能通过控制交换机的每一个端口来控制网络用户对网络资源的访问,同时 VLAN 和第三层、第四层的交换结合使用能够为网络提供较好的安全措施。

VLAN 是为解决以太网的广播问题和安全性而提出的,它在以太网帧的基础上增加了

VLAN 头,用 VLAN ID 把用户划分为更小的工作组,限制不同工作组间的用户两层互访。虚拟局域网的好处是可以限制广播范围,并能够形成虚拟工作组,动态管理网络。

既然 VLAN 隔离了广播风暴,同时也隔离了各个不同的 VLAN 之间的通信,所以不同的 VLAN 之间的通信是需要由路由来完成的。

8.1.2 VLAN 的优点

(1) 改进管理效率。VLAN 提供了有效的机制来控制由于企业结构、人事或资源变化对网络系统所造成的影响。例如,当用户从一个地点移到另一个地点时,只需对交换机的配置稍做改动即可,大大简化了重新布线、配置和测试的步骤。

(2) 控制广播活动。VLAN 可保护网络不受由广播流量所造成的影响,一个 VLAN 内的广播信息不会传送到 VLAN 之外,网络管理人员可以通过设置 VLAN 灵活控制广播域。

(3) 增强网络安全。VLAN 创造了虚拟边界,它只能通过路由跨越,因此通过将网络用户划分到不同 VLAN 并结合访问控制,可控制 VLAN 外部站点对 VLAN 内部资源的访问,提高网络的安全性。

8.1.3 VLAN 划分

(1) 根据端口来划分 VLAN。利用交换机的端口来划分 VLAN 成员。被设定的端口都在同一个广播域中。例如,一个交换机的 1~5 端口被定义为虚拟网 AAA,同一交换机的 6~8 端口组成虚拟网 BBB。这样做允许各端口之间的通信,并允许共享型网络的升级。但是,这种划分模式将虚拟网限制在了一台交换机上。

(2) 根据 MAC 地址划分 VLAN。根据每个主机的 MAC 地址来划分,即对每个 MAC 地址的主机都配置它属于哪个组。最大优点就是当用户物理位置移动时,即从一个交换机换到其他的交换机时,VLAN 不用重新配置,所以,可以认为这种根据 MAC 地址的划分方法是基于用户的 VLAN。缺点是初始化时,所有的用户都必须进行配置,如果有几百个甚至上千个用户,配置是非常累的。

(3) 根据网络层划分 VLAN。根据每个主机的网络层地址或协议类型(如果支持多协议)划分,虽然这种划分方法是根据网络地址,比如 IP 地址,但它不是路由,与网络层的路由毫无关系。优点是用户的物理位置改变了,不需要重新配置所属的 VLAN,而且可以根据协议类型来划分 VLAN,这对网络管理者来说很重要。还有,这种方法不需要附加的帧标签来识别 VLAN,这样可以减少网络的通信量。

(4) 根据 IP 组播划分 VLAN。IP 组播实际上也是一种 VLAN 的定义,即认为一个组播组就是一个 VLAN,这种划分的方法将 VLAN 扩大到了广域网,因此这种方法具有更大的灵活性,而且也很容易通过路由器进行扩展,当然这种方法不适合局域网,主要是效率不高。

(5) 基于规则的 VLAN。也称为基于策略的 VLAN。这是最灵活的 VLAN 划分方法,具有自动配置的能力,能够把相关的用户连成一体,在逻辑划分上称为"关系网络"。网络管理员只需在网管软件中确定划分 VLAN 的规则(或属性),那么当一个站点加入网络中时,将会被"感知",并被自动地包含进正确的 VLAN 中。同时,对站点的移动和改变也可自动识别和跟踪。

8.1.4 VLAN 标准

（1）802.10VLAN 标准。在 1995 年，Cisco 公司提倡使用 IEEE 802.10 协议。然而，大多数 802 委员会的成员都反对推广 802.10。

（2）802.1Q。在 1996 年 3 月，IEEE 802.1 Internetworking 委员会出台的标准进一步完善了 VLAN 的体系结构，统一了不同厂商的标签格式，并制定了 VLAN 标准在未来的发展方向，形成的 802.1Q 的标准在业界获得了广泛的推广。

（3）Cisco ISL(Inter-Switch Link，ISL)标签是 Cisco 公司的专有封装方式，因此只能在 Cisco 的设备上支持。ISL 是一个在交换机之间、交换机与路由器之间及交换机与服务器之间传递多个 VLAN 信息及 VLAN 数据流的协议，通过直接在交换机的端口配置 ISL 封装，即可跨越交换机进行整个网络的 VLAN 分配和配置。

8.2 实验 1 静态 VLAN

8.2.1 实验拓扑

PC1 和 PC2 分别连接 Switch 交换机的快速以太网端口 Fa0/1 和 Fa0/2，见图 8.2。

设备名	IP 地址	MAC 地址
PC1	192.168.1.1	0009.7C70.C10E
PC2	192.168.1.2	0050.0F81.2830

8.2.2 实验内容

（1）配置交换机 VLAN 信息。
（2）配置端口 VLAN 信息。

8.2.3 实验步骤

（1）查看初始 VLAN 信息。

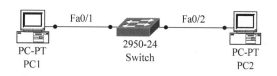

图 8.2 静态 VLAN 配置拓扑图

```
Switch > enable
Switch # show vlan

VLAN Name                             Status    Ports
---- -------------------------------- --------- -------------------------------
1    default                          active    Fa0/1, Fa0/2, Fa0/3, Fa0/4
                                                Fa0/5, Fa0/6, Fa0/7, Fa0/8
                                                Fa0/9, Fa0/10, Fa0/11, Fa0/12
                                                Fa0/13, Fa0/14, Fa0/15, Fa0/16
                                                Fa0/17, Fa0/18, Fa0/19, Fa0/20
                                                Fa0/21, Fa0/22, Fa0/23, Fa0/24
1002 fddi-default                     act/unsup
1003 token-ring-default               act/unsup
1004 fddinet-default                  act/unsup
1005 trnet-default                    act/unsup
```

```
VLAN  Type   SAID    MTU   Parent  RingNo  BridgeNo  Stp   BrdgMode  Trans1  Trans2
----  ----   ------  ----  ------  ------  --------  ----  --------  ------  ------
1     enet   100001  1500  -       -       -         -     -         0       0
1002  fddi   101002  1500  -       -       -         -     -         0       0
1003  tr     101003  1500  -       -       -         -     -         0       0
1004  fdnet  101004  1500  -       -       -         ieee  -         0       0
1005  trnet  101005  1500  -       -       -         ibm   -         0       0

-- More --
```

【提示】 Fa0/1 端口和 Fa0/2 端口都属于 VLAN1。

(2) 创建 VLAN2, 查看 VLAN 信息。

```
Switch#configure terminal
Enter configuration commands, one per line. End with CNTL/Z.
Switch(config)#vlan 2
Switch(config-vlan)#name VLAN2
Switch(config-vlan)#end
Switch#
%SYS-5-CONFIG_I: Configured from console by console

Switch#show vlan

VLAN  Name                    Status      Ports
----  ----------------------  ---------   -------------------------------
1     default                 active      Fa0/1, Fa0/2, Fa0/3, Fa0/4
                                          Fa0/5, Fa0/6, Fa0/7, Fa0/8
                                          Fa0/9, Fa0/10, Fa0/11, Fa0/12
                                          Fa0/13, Fa0/14, Fa0/15, Fa0/16
                                          Fa0/17, Fa0/18, Fa0/19, Fa0/20
                                          Fa0/21, Fa0/22, Fa0/23, Fa0/24
2     VLAN2                   active
1002  fddi-default            act/unsup
1003  token-ring-default      act/unsup
1004  fddinet-default         act/unsup
1005  trnet-default           act/unsup

VLAN  Type   SAID    MTU   Parent  RingNo  BridgeNo  Stp   BrdgMode  Trans1  Trans2
----  ----   ------  ----  ------  ------  --------  ----  --------  ------  ------
1     enet   100001  1500  -       -       -         -     -         0       0
2     enet   100002  1500  -       -       -         -     -         0       0
1002  fddi   101002  1500  -       -       -         -     -         0       0
1003  tr     101003  1500  -       -       -         -     -         0       0
1004  fdnet  101004  1500  -       -       -         ieee  -         0       0
-- More --
```

【提示】 VLAN 信息中增加了 VLAN2 信息, 但是没有端口属于 VLAN2。

(3) PC1 ping PC2。

```
PC>ping 192.168.1.2
```

```
Pinging 192.168.1.2 with 32 bytes of data:

Reply from 192.168.1.2: bytes = 32 time = 2ms TTL = 128
Reply from 192.168.1.2: bytes = 32 time = 2ms TTL = 128
Reply from 192.168.1.2: bytes = 32 time = 1ms TTL = 128
Reply from 192.168.1.2: bytes = 32 time = 3ms TTL = 128
```

```
Ping statistics for 192.168.1.2:
    Packets: Sent = 4, Received = 4, Lost = 0 (0% loss),
Approximate round trip times in milli-seconds:
Minimum = 1ms, Maximum = 3ms, Average = 2ms
```

【提示】 由于 Fa0/1 端口和 Fa0/2 端口都属于 VLAN1,所以连接属于 VLAN1 端口的 PC1 能够 ping 通连接同属于 VLAN1 端口的 PC2。

(4) 分配 Fa0/2 端口到 VLAN2,查看 VLAN 信息。

```
Switch#configure terminal
Enter configuration commands, one per line. End with CNTL/Z.
Switch(config)#interface fastEthernet 0/2
Switch(config-if)#switchport mode access
Switch(config-if)#switchport access vlan 2
Switch(config-if)#end
Switch#
%SYS-5-CONFIG_I: Configured from console by console
Switch#show vlan

VLAN Name                             Status    Ports
---- -------------------------------- --------- -------------------------------
1    default                          active    Fa0/1, Fa0/3, Fa0/4, Fa0/5
                                                Fa0/6, Fa0/7, Fa0/8, Fa0/9
                                                Fa0/10, Fa0/11, Fa0/12, Fa0/13
                                                Fa0/14, Fa0/15, Fa0/16, Fa0/17
                                                Fa0/18, Fa0/19, Fa0/20, Fa0/21
                                                Fa0/22, Fa0/23, Fa0/24
2    VLAN2                            active    Fa0/2
1002 fddi-default                     act/unsup
1003 token-ring-default               act/unsup
1004 fddinet-default                  act/unsup
1005 trnet-default                    act/unsup

VLAN Type  SAID    MTU  Parent RingNo BridgeNo Stp  BrdgMode Trans1 Trans2
---- ----- ------- ---- ------ ------ -------- ---- -------- ------ ------
1    enet  100001  1500 -      -      -        -    -        0      0
2    enet  100002  1500 -      -      -        -    -        0      0
1002 fddi  101002  1500 -      -      -        -    -        0      0
1003 tr    101003  1500 -      -      -        -    -        0      0
1004 fdnet 101004  1500 -      -      -        ieee -        0      0
--More--
```

【提示】 Fa0/2 端口属于 VLAN2，Fa0/1 端口仍属于 VLAN1。

（5）PC1 再次 ping PC2。

```
PC>ping 192.168.1.2

Pinging 192.168.1.2 with 32 bytes of data:

Request timed out.
Request timed out.
Request timed out.
Request timed out.

Ping statistics for 192.168.1.2:
Packets: Sent = 4, Received = 0, Lost = 4 (100% loss),
```

【提示】 由于 Fa0/1 端口和 Fa0/2 端口属于不同的 VLAN，所以连接属于 VLAN1 端口的 PC1 不会 ping 通连接属于 VLAN2 端口的 PC2。为了使 PC1 与 PC2 能够通信，一种方法是配置 PC1 与 PC2 属于同一个 VLAN；一种方法是借助单臂路由或者三层交换实现 VLAN 间路由。

8.3 实验 2 干道链接

8.3.1 实验拓扑

PC1 和 PC2 分别连接 Switch0 交换机的快速以太网端口 Fa0/2 和 Fa0/3，PC3 和 PC4 分别连接 Switch1 交换机快速以太网端口 Fa0/2 和 Fa0/3，Switch0 交换机的快速以太网端口 Fa0/1 连接 Switch1 交换机的快速以太网端口 Fa0/1，见图 8.3。

设备名	IP 地址
PC1	192.168.1.1
PC2	192.168.1.2
PC3	192.168.1.3
PC4	192.168.1.4

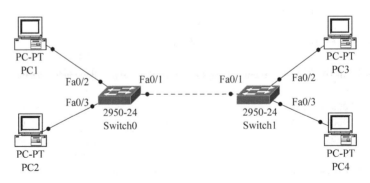

图 8.3 干道链接拓扑

8.3.2 实验内容

（1）配置交换机端口为干道端口。
（2）配置干道端口本征 VLAN。
（3）配置干道端口 VLAN 允许列表。

8.3.3 实验步骤

（1）交换机默认配置下，测试 PC 之间连通性。

```
PC1：
PC > ping 192.168.1.2

Pinging 192.168.1.2 with 32 bytes of data:

Reply from 192.168.1.2: bytes = 32 time = 25ms TTL = 128
Reply from 192.168.1.2: bytes = 32 time = 12ms TTL = 128
Reply from 192.168.1.2: bytes = 32 time = 11ms TTL = 128
Reply from 192.168.1.2: bytes = 32 time = 14ms TTL = 128

Ping statistics for 192.168.1.2:
    Packets: Sent = 4, Received = 4, Lost = 0 (0 % loss),
Approximate round trip times in milli - seconds:
    Minimum = 11ms, Maximum = 25ms, Average = 15ms
```

【提示】 Switch0 的快速以太网端口 Fa0/2 和 Fa0/3 默认配置属于 VLAN1 的成员，所以 PC1 ping PC2 属于同一个 VLAN 通信。

```
PC > ping 192.168.1.3

Pinging 192.168.1.3 with 32 bytes of data:

Reply from 192.168.1.3: bytes = 32 time = 25ms TTL = 128
Reply from 192.168.1.3: bytes = 32 time = 12ms TTL = 128
Reply from 192.168.1.3: bytes = 32 time = 11ms TTL = 128
Reply from 192.168.1.3: bytes = 32 time = 14ms TTL = 128

Ping statistics for 192.168.1.3:
    Packets: Sent = 4, Received = 4, Lost = 0 (0 % loss),
Approximate round trip times in milli - seconds:
    Minimum = 11ms, Maximum = 25ms, Average = 15ms
```

```
PC > ping 192.168.1.4

Pinging 192.168.1.4 with 32 bytes of data:

Reply from 192.168.1.4: bytes = 32 time = 25ms TTL = 128
Reply from 192.168.1.4: bytes = 32 time = 12ms TTL = 128
```

```
Reply from 192.168.1.4: bytes = 32 time = 11ms TTL = 128
Reply from 192.168.1.4: bytes = 32 time = 14ms TTL = 128

Ping statistics for 192.168.1.4:
    Packets: Sent = 4, Received = 4, Lost = 0 (0% loss),
Approximate round trip times in milli - seconds:
    Minimum = 11ms, Maximum = 25ms, Average = 15ms
```

【提示】 Switch0 和 Switch1 的快速以太网端口 Fa0/2、Fa0/3 和 Fa0/1 默认配置都属于 VLAN1 的成员，所以 PC1 ping PC3 和 PC1 ping PC4 属于同一个 VLAN 通信。

（2）在 Switch0 和 Switch1 上创建 VLAN2，并配置各自的快速以太网端口 Fa0/3 属于 VLAN2。

```
Switch0:
Switch0#configure terminal
Enter configuration commands, one per line. End with CNTL/Z.
Switch0(config)#vlan 2
Switch0(config-vlan)#end
Switch0#
%SYS-5-CONFIG_I: Configured from console by console

Switch0#config terminal
Enter configuration commands, one per line. End with CNTL/Z.
Switch0(config)#interface fastEthernet 0/1
Switch0(config-if)#switchport access vlan 2
Switch0(config-if)#end
Switch0#
%SYS-5-CONFIG_I: Configured from console by console
```

```
Switch1:
Switch1#configure terminal
Enter configuration commands, one per line. End with CNTL/Z.
Switch1(config)#vlan 2
Switch1(config-vlan)#end
Switch1#
%SYS-5-CONFIG_I: Configured from console by console

Switch1#config terminal
Enter configuration commands, one per line. End with CNTL/Z.
Switch1(config)#interface fastEthernet 0/1
Switch1(config-if)#switchport access vlan 2
Switch1(config-if)#end
Switch1#
%SYS-5-CONFIG_I: Configured from console by console
```

（3）再次测试 PC 之间的连通性。

```
PC1:
PC>ping 192.168.1.2
```

```
Pinging 192.168.1.2 with 32 bytes of data:

Request timed out.
Request timed out.
Request timed out.
Request timed out.

Ping statistics for 192.168.1.2:
    Packets: Sent = 4, Received = 0, Lost = 4 (100% loss),
```

【提示】 Switch0 的快速以太网端口 Fa0/2 属于 VLAN1 的成员，Fa0/3 属于 VLAN2 的成员，所以 PC1 ping PC2 不属于同一个 VLAN 通信。

```
PC > ping 192.168.1.3

Pinging 192.168.1.3 with 32 bytes of data:

Reply from 192.168.1.3: bytes = 32 time = 25ms TTL = 128
Reply from 192.168.1.3: bytes = 32 time = 12ms TTL = 128
Reply from 192.168.1.3: bytes = 32 time = 11ms TTL = 128
Reply from 192.168.1.3: bytes = 32 time = 14ms TTL = 128

Ping statistics for 192.168.1.3:
    Packets: Sent = 4, Received = 4, Lost = 0 (0% loss),
Approximate round trip times in milli - seconds:
    Minimum = 11ms, Maximum = 25ms, Average = 15ms
```

【提示】 Switch0 和 Switch1 的快速以太网端口 Fa0/2 和 Fa0/1 都属于 VLAN1 的成员，所以 PC1 ping PC3 属于同一个 VLAN 通信。

```
PC > ping 192.168.1.4

Pinging 192.168.1.4 with 32 bytes of data:

Request timed out.
Request timed out.
Request timed out.
Request timed out.

Ping statistics for 192.168.1.4:
    Packets: Sent = 4, Received = 0, Lost = 4 (100% loss),
```

【提示】 虽然 Switch0 和 Switch1 的快速以太网端口 Fa0/2 都属于 VLAN2 的成员，但是快速以太网端口 Fa0/1 都属于 VLAN1 的成员，所以虽然 PC1 和 PC4 属于同一个 VLAN，但是 ping 要进行 VLAN 间通信，即要通过 VLAN1 进行通信，但是 VLAN 间通信需要第三层设备转发。

(4) Switch0 和 Switch1 上配置各自的快速以太网端口 Fa0/1 为干道端口。

```
Switch0:
Switch0 > enable
Switch0 # configure terminal
Enter configuration commands, one per line. End with CNTL/Z.
Switch0(config)# interface fastEthernet 0/1
Switch0(config-if)# switchport mode trunk
Switch0(config-if)# end
Switch0 #
% SYS-5-CONFIG_I: Configured from console by console
```

```
Switch1:
Switch1 > enable
Switch1 # configure terminal
Enter configuration commands, one per line. End with CNTL/Z.
Switch1(config)# interface fastEthernet 0/1
Switch1(config-if)# switchport mode trunk
Switch1(config-if)# end
Switch0 #
% SYS-5-CONFIG_I: Configured from console by console
```

【提示】 端口模式设置选项包括 trunk、dynamic desirable 和 dynamic auto。

trunk：端口为永久干道模式。

dynamic desirable：端口主动变为干道模式。如果连接另一端的端口为 negotiate，dynamic desirable 或者 dynamic auto，则协商成功，端口变为干道模式，否则协商不成功，不能变为干道模式。

dynamic auto：端口被动变为干道模式。如果连接另一端的端口为 negotiate 或者 dynamic desirable，则协商成功，端口变为干道模式，否则协商不成功，不能变为干道模式。

(5) 再次测试 PC 之间的连通性。

```
PC1:
PC > ping 192.168.1.2

Pinging 192.168.1.2 with 32 bytes of data:

Request timed out.
Request timed out.
Request timed out.
Request timed out.

Ping statistics for 192.168.1.2:
    Packets: Sent = 4, Received = 0, Lost = 4 (100% loss),
```

【提示】 Switch0 的快速以太网端口 Fa0/2 属于 VLAN1 的成员，Fa0/3 属于 VLAN2 的成员，所以 PC1 ping PC2 不属于同一个 VLAN 通信。

```
PC > ping 192.168.1.3

Pinging 192.168.1.3 with 32 bytes of data:

Reply from 192.168.1.3: bytes = 32 time = 25ms TTL = 128
Reply from 192.168.1.3: bytes = 32 time = 12ms TTL = 128
Reply from 192.168.1.3: bytes = 32 time = 11ms TTL = 128
Reply from 192.168.1.3: bytes = 32 time = 14ms TTL = 128

Ping statistics for 192.168.1.3:
    Packets: Sent = 4, Received = 4, Lost = 0 (0% loss),
Approximate round trip times in milli - seconds:
    Minimum = 11ms, Maximum = 25ms, Average = 15ms
```

【提示】 Switch0 和 Switch1 的快速以太网端口 Fa0/2 都属于 VLAN1 的成员，快速以太网端口 Fa0/1 是干道端口，属于所有 VLAN 的成员，所以 PC1 ping PC3 属于同一个 VLAN 通信。

```
PC > ping 192.168.1.4

Pinging 192.168.1.4 with 32 bytes of data:

Request timed out.
Request timed out.
Request timed out.
Request timed out.

Ping statistics for 192.168.1.4:
    Packets: Sent = 4, Received = 0, Lost = 4 (100% loss),
```

【提示】 虽然 Switch0 和 Switch1 的快速以太网端口 Fa0/1 是干道端口，属于所有 VLAN 的成员，但是 Switch0 的快速以太网端口 Fa0/2 属于 VLAN1 成员，Switch1 的快速以太网端口 Fa0/3 属于 VLAN2 成员，所以 PC1 ping PC4 不属于同一个 VLAN 通信。

(6) 配置干道端口 Fa0/1 本征 VLAN 为 VLAN2。

```
Switch0 # configure terminal
Enter configuration commands, one per line. End with CNTL/Z.
Switch0(config) # interface fastEthernet 0/1
Switch0(config - if) # switchport trunk native vlan 2
Switch0(config - if) # end
Switch0 #
% SYS - 5 - CONFIG_I: Configured from console by console
```

【提示】 在干道链路上，数据帧会根据 ISL 或者 802.1Q 被重新封装，然而如果是本征 VLAN 的数据，是不会被重新封装而就在干道链路上传输。所以要求干道链路两端的本征 VLAN 要一致，否则，交换机会提示出错。

(7) 干道端口 Fa0/1 的 VLAN 允许列表移除 VLAN 1。

```
Switch0 # configure terminal
Enter configuration commands, one per line. End with CNTL/Z.
Switch0(config) # interface fastEthernet 0/1
Switch0(config-if) # switchport trunk allowed vlan remove 1
Switch0(config-if) # end
Switch0 #
% SYS-5-CONFIG_I: Configured from console by console
```

【提示】 VLAN 列表的格式是 all|add|remove|except $vlan\text{-}list[,vlan-list\cdots]$
all：允许所有 VLAN。
add：添加指定 VLAN 到当前的 VLAN 允许列表。
remove：从当前 VLAN 允许列表移除指定 VLAN。
except：允许除指定 VLAN 外的所有 VLAN。
$vlan\text{-}list$：单独 VLAN 或者连续 VLAN 范围。

(8) 查看干道端口 Fa0/1 运行配置。

```
Switch0 # show running-config
Building configuration...

Current configuration : 1115 bytes
!
interface FastEthernet0/1
 switchport trunk native vlan 2
 switchport trunk allowed vlan 2-1005
 switchport mode trunk
!
```

【提示】 IOS 自动根据 VLAN 允许列表设置调整干道端口相关的运行配置。

(9) 测试 PC1 与 PC3、PC2 与 PC4 的连通性。

```
PC1:

PC > ping 192.168.1.3

Pinging 192.168.1.3 with 32 bytes of data:

Request timed out.
Request timed out.
Request timed out.
Request timed out.

Ping statistics for 192.168.1.3:
Packets: Sent = 4, Received = 0, Lost = 4 (100% loss),
```

```
PC2:
    PC > ping 192.168.1.4

Pinging 192.168.1.4 with 32 bytes of data:

Reply from 192.168.1.4: bytes = 32 time = 13ms TTL = 128
Reply from 192.168.1.4: bytes = 32 time = 14ms TTL = 128
Reply from 192.168.1.4: bytes = 32 time = 13ms TTL = 128
Reply from 192.168.1.4: bytes = 32 time = 15ms TTL = 128

Ping statistics for 192.168.1.4:
    Packets: Sent = 4, Received = 4, Lost = 0 (0% loss),
Approximate round trip times in milli - seconds:
Minimum = 13ms, Maximum = 15ms, Average = 13ms
```

【提示】 由于 PC1 ping PC3 属于 VLAN1 通信,但是 VLAN1 不在干道端口 Fa0/1 的 VLAN 允许列表里,即不允许 VLAN1 通信经过该干道端口,所以 PC1 不能 ping 通 PC3。由于 PC2 ping PC4 属于 VLAN2 通信,而且 VLAN2 在干道端口 Fa0/1 的 VLAN 允许列表里,即允许 VLAN2 通信经过该干道端口,所以 PC2 能够 ping 通 PC4。

(10) 查看干道端口 Fa0/1 信息。

```
Switch0 # show interfaces fa0/1 switchport
Name: Fa0/1
Switchport: Enabled
Administrative Mode: trunk
Operational Mode: trunk
Administrative Trunking Encapsulation: dot1q
Operational Trunking Encapsulation: dot1q
Negotiation of Trunking: On
Access Mode VLAN: 1 (default)
Trunking Native Mode VLAN: 1 (default)
Voice VLAN: none
Administrative private - vlan host - association: none
Administrative private - vlan mapping: none
Administrative private - vlan trunk native VLAN: none
Administrative private - vlan trunk encapsulation: dot1q
Administrative private - vlan trunk normal VLANs: none
Administrative private - vlan trunk private VLANs: none
Operational private - vlan: none
Trunking VLANs Enabled: ALL
Pruning VLANs Enabled: 2 - 1001
Capture Mode Disabled
Capture VLANs Allowed: ALL
Protected: false
Appliance trust: none
```

8.4 实验 3　干道协议

8.4.1 实验拓扑

交换机 Transparent 通过干道端口 Fa0/1 和 Fa0/2 分别与交换机 Server 和交换机 Client 各自的干道端口 Fa0/1 连接，见图 8.4。

图 8.4　干道协议配置

8.4.2 实验内容

(1) 配置 VTP 版本。
(2) 配置 VTP 域名和域密码。
(3) 配置 VTP 模式。
(4) 配置 VTP 修剪。

8.4.3 实验步骤

(1) 配置交换机 VTP 版本为 2。

```
Server:
Server#config terminal
Enter configuration commands, one per line.  End with CNTL/Z.
Server(config)#vtp version 2
Server(config)#
```

```
Transparent:
Transparent#config terminal
Enter configuration commands, one per line.  End with CNTL/Z.
Transparent(config)#vtp version 2
Transparent(config)#
```

```
Client:
Client#configure terminal
Enter configuration commands, one per line.  End with CNTL/Z.
Client(config)#vtp version 2
Client(config)#
```

【提示】　在管理域中可以运行两种不同版本的 VTP：VTP 版本 1 和 VTP 版本 2，由于两个版本的 VTP 不能互相操作，所以管理域中的所有交换机的 VTP 版本要一致。

（2）配置 VTP 域名为 test，域密码为 cisco。

```
Server:
Server(config)# vtp domain test
Changing VTP domain name from NULL to test
Server(config)# vtp password cisco
Setting device VLAN database password to cisco
Server(config)#
```

```
Transparent:
Transparent(config)# vtp domain test
Changing VTP domain name from NULL to test
Transparent(config)# vtp password cisco
Setting device VLAN database password to cisco
Transparent(config)#
```

```
Client:
Client(config)# vtp domain test
Changing VTP domain name from NULL to test
Client(config)# vtp password cisco
Setting device VLAN database password to cisco
Client(config)#
```

【提示】 同一个管理域的所有交换机的 VTP 域名和域密码要一致。

（3）配置 VTP 模式，Server 为服务器模式，Client 为客户机模式，Transparent 为透明模式。

```
Server:
Server(config)# vtp mode server
Device mode already VTP SERVER.
```

```
Client:
Client(config)# vtp mode client
Setting device to VTP CLIENT mode.
```

```
Transparent:
    Transparent(config)# vtp mode transparent
Setting device to VTP TRANSPARENT mode.
```

【提示】 VTP 模式包括服务器（server）模式、客户机（client）模式和透明（transparent）模式。

服务器（server）模式：所有交换机默认使用的是 VTP 服务器模式。在 VTP 域中至少需要一台服务器来传播 VLAN 信息。服务器模式必须完成的内容：

① 在 VTP 域中建立、添加和删除 VLAN；改变 VTP 信息。

② 改变 VTP 信息。对于服务器模式下交换机的改变被广播到整个 VTP 域中。

全局 VLAN 必须在服务器上进行配置。服务器将向交换机配置中添加 VLAN，这样每次交换机启动时，VLAN 的信息将会进行传播。

客户机(client)模式：VTP 客户机接收来自 VTP 服务器的信息，并且收发更新信息，但是只要它们还是客户机，它们就不能对 VTP 配置进行任何修改。在 VTP 服务器通知客户机交换及有关新的 VLAN 信息之前，客户机交换机上的端口是不会被添加到新的 VLAN 之中的。如果希望交换机成为服务器，首先要使它成为客户机，这样它就可以收到所有正确的 VLAN 信息，然后再把它改成服务器。如果交换机断电了，那么它将不会保留任何全局 VTP 的信息。

透明(transparent)模式：VTP 透明模式交换机并不参与到 VTP 域中，但是它们将依然通过已配置好的干道连接接收和转发 VTP 通知信息。然而，对于透明模式的交换机而言，要想将 VLAN 的信息通过已配置的干道链路广播出去，必须使用版本 2 的 VTP。如果不这样，交换机将不会转发任何信息。由于 VTP 透明模式交换机保留自己的数据库并且不与其他的交换机共享这个数据库，因此它可以添加和删除 VLAN。透明模式交换机被认为是本地意义上的设备。

(4) Server 上创建 VLAN2 和 VLAN3，Transparent 上创建 VLAN4，Client 上查看 VLAN 信息。

```
Server:
    Server(config)#vlan 2
    Server(config-vlan)#exit
Server(config)#vlan 3
Server(config-vlan)#

Transparent:
    Transparent(config)#vlan 4
Transparent(config-vlan)#end
Transparent#
%SYS-5-CONFIG_I: Configured from console by console

Transparent#show vlan

VLAN Name                             Status     Ports
---- -------------------------------- ---------- -------------------------------
1    default                          active     Fa0/3, Fa0/4, Fa0/5, Fa0/6
                                                 Fa0/7, Fa0/8, Fa0/9, Fa0/10
                                                 Fa0/11, Fa0/12, Fa0/13, Fa0/14
                                                 Fa0/15, Fa0/16, Fa0/17, Fa0/18
                                                 Fa0/19, Fa0/20, Fa0/21, Fa0/22
                                                 Fa0/23, Fa0/24
4    VLAN0004                         active
1002 fddi-default                     act/unsup
1003 token-ring-default               act/unsup
1004 fddinet-default                  act/unsup
1005 trnet-default                    act/unsup
```

```
VLAN  Type   SAID    MTU   Parent  RingNo  BridgeNo  Stp   BrdgMode  Trans1  Trans2
---   ----   -----   ---   ------  ------  --------  ----  --------  ------  ------
1     enet   100001  1500  -       -       -         -     -         0       0
1002  fddi   101002  1500  -       -       -         -     -         0       0
1003  tr     101003  1500  -       -       -         -     -         0       0
1004  fdnet  101004  1500  -       -       -         ieee  -         0       0
1005  trnet  101005  1500  -       -       -         ibm   -         0       0

-- More --
```

Client:
```
    Client#show vlan

VLAN  Name                  Status      Ports
---   -------------------   ---------   -------------------------------
1     default               active      Fa0/2, Fa0/3, Fa0/4, Fa0/5
                                        Fa0/6, Fa0/7, Fa0/8, Fa0/9
                                        Fa0/10, Fa0/11, Fa0/12, Fa0/13
                                        Fa0/14, Fa0/15, Fa0/16, Fa0/17
                                        Fa0/18, Fa0/19, Fa0/20, Fa0/21
                                        Fa0/22, Fa0/23, Fa0/24
2     VLAN0002              active
3     VLAN0003              active
1002  fddi-default          act/unsup
1003  token-ring-default    act/unsup
1004  fddinet-default       act/unsup
1005  trnet-default         act/unsup

VLAN  Type  SAID    MTU   Parent  RingNo  BridgeNo  Stp   BrdgMode  Trans1  Trans2
---   ----  -----   ---   ------  ------  --------  ----  --------  ------  ------
1     enet  100001  1500  -       -       -         -     -         0       0
2     enet  100002  1500  -       -       -         -     -         0       0
3     enet  100003  1500  -       -       -         -     -         0       0
1002  fddi  101002  1500  -       -       -         -     -         0       0

-- More --
```

【提示】 由于 Transparent 是透明模式，仅转发来自 Server（服务器模式）的 VTP 信息，不根据转发的 VTP 信息更新其 VLAN 信息，只根据自身信息更新其 VLAN 信息，所以没有看到 VLAN2 和 VLAN3 信息，可以看到 VLAN4 信息。

由于 Client 是客户机模式，同时 Transparent 不发送自身的 VTP 信息，所以只能接收从 Transparent 转发的 Server 的 VTP 信息，更新其 VLAN 信息，所以可以看到 VLAN2 和 VLAN3 信息，看不到 VLAN4 信息。

（5）查看 VTP 统计信息。

Server:
```
    Server(config-vlan)#end
```

```
Server#
% SYS - 5 - CONFIG_I: Configured from console by console

Server# show vtp status
VTP Version                        : 2
Configuration Revision             : 7
Maximum VLANs supported locally    : 255
Number of existing VLANs           : 7
VTP Operating Mode                 : Server
VTP Domain Name                    : test
VTP Pruning Mode                   : Disabled
VTP V2 Mode                        : Disabled
VTP Traps Generation               : Disabled
MD5 digest                         : 0xC6 0xE6 0xCE 0x6B 0xA5 0x22 0xA5 0x07
Configuration last modified by 0.0.0.0 at 3 - 1 - 93 00:52:20
Local updater ID is 0.0.0.0 (no valid interface found)
Server#
```

```
Transparent:
    Transparent# show vtp status
VTP Versio                         : 2
Configuration Revision             : 0
Maximum VLANs supported locally    : 255
Number of existing VLANs           : 5
VTP Operating Mode                 : Transparent
VTP Domain Name                    : test
VTP Pruning Mode                   : Disabled
VTP V2 Mode                        : Enabled
VTP Traps Generation               : Disabled
MD5 digest                         : 0x3B 0x21 0xDF 0x53 0x8C 0x6C 0xC4 0x46
Configuration last modified by 0.0.0.0 at 3 - 1 - 93 00:13:39
Transparent#
```

```
Client:
    Client# show vtp status
VTP Version                        : 2
Configuration Revision             : 7
Maximum VLANs supported locally    : 255
Number of existing VLANs           : 7
VTP Operating Mode                 : Client
VTP Domain Name                    : test
VTP Pruning Mode                   : Disabled
VTP V2 Mode                        : Disabled
VTP Traps Generation               : Disabled
MD5 digest                         : 0xC6 0xE6 0xCE 0x6B 0xA5 0x22 0xA5 0x07
Configuration last modified by 0.0.0.0 at 3 - 1 - 93 00:52:20
Client#
```

(6) Server 删除 VLAN3,再次查看 VTP 统计信息。

```
Server:
    Server#config terminal
Enter configuration commands, one per line. End with CNTL/Z.
Server(config)#no vlan 3
Server(config)#end
Server#
%SYS-5-CONFIG_I: Configured from console by console

Server#show vtp status
VTP Version                      : 2
Configuration Revision           : 8
Maximum VLANs supported locally  : 255
Number of existing VLANs         : 6
VTP Operating Mode               : Server
VTP Domain Name                  : test
VTP Pruning Mode                 : Disabled
VTP V2 Mode                      : Disabled
VTP Traps Generation             : Disabled
MD5 digest                       : 0x66 0x6A 0xD0 0x58 0x21 0x6C 0x2D 0xB3
Configuration last modified by 0.0.0.0 at 3-1-93 01:07:49
Local updater ID is 0.0.0.0 (no valid interface found)
Server#
```

```
Transparent:
    Transparent#show vtp status
VTP Version                      : 2
Configuration Revision           : 0
Maximum VLANs supported locally  : 255
Number of existing VLANs         : 6
VTP Operating Mode               : Transparent
VTP Domain Name                  : test
VTP Pruning Mode                 : Disabled
VTP V2 Mode                      : Disabled
VTP Traps Generation             : Disabled
MD5 digest                       : 0xB6 0x3A 0xF3 0x3C 0x09 0x80 0x91 0x94
Configuration last modified by 0.0.0.0 at 3-1-93 00:35:34
Transparent#
```

```
Client:
    Client#show vtp status
VTP Version                      : 2
Configuration Revision           : 8
Maximum VLANs supported locally  : 255
Number of existing VLANs         : 6
VTP Operating Mode               : Client
VTP Domain Name                  : test
VTP Pruning Mode                 : Disabled
```

```
VTP V2 Mode                          : Disabled
VTP Traps Generation                 : Disabled
MD5 digest                           : 0x66 0x6A 0xD0 0x58 0x21 0x6C 0x2D 0xB3
Configuration last modified by 0.0.0.0 at 3-1-93 01:07:49
Client#
```

【提示】 Client(客户机模式)的 VTP 配置修订号随 Server(服务器模式)的 VTP 配置修订号更新,而 Transparent(透明模式)的 VTP 配置修订号不随 Server(服务器模式)的 VTP 配置修订号更新。

(7) Server 上启用 VTP 修剪。

```
    Server#configure terminal
Enter configuration commands, one per line. End with CNTL/Z
    Server(config)#vtp pruning
```

(8) Server 干道端口 Fa0/1 的 VLAN 剪裁列表移除 VLAN4。

```
Server(config)#interface fastEthernet 0/1
Server(config-if)#switchport trunk pruning vlan remove 4
```

【提示】 VLAN 列表格式是 all|add|remove|except *vlan-list*[,*vlan-list*…]
all:允许所有 VLAN。
add:添加指定 VLAN 到当前的 VLAN 剪裁列表。
remove:从当前 VLAN 剪裁列表移除指定 VLAN。
except:允许除指定 VLAN 外的所有 VLAN。
vlan-list:单独 VLAN 或者连续 VLAN 范围。

第 9 章　生　成　树

9.1　基 础 知 识

9.1.1　STP 概述

生成树协议(Spanning Tree Protocol,STP)是一种第二层的链路管理协议,它用于维护一个无环路的网络。IEEE 802 委员会在 IEEE 802.1d 规范中公布。STP 的目的是为了维护一个无环路的网络拓扑。当交换机和网桥在拓扑中发现环路时,它们自动地在逻辑上阻塞一个或多个冗余端口,从而获得无环路的拓扑。STP 持续探测网络,以便在链路、交换机或网桥失效或增加时都可以获得响应。在网络拓扑改变时,运行 STP 的交换机和网桥自动重新配置它们的端口,以避免失去连接或者产生环路。

在基于交换机的互联网络中,物理层的环路会导致严重的问题。广播风暴、帧的重复传送以及 MAC 数据库的不稳定都会导致网络瘫痪。交换式网络使用 STP 在由环路的物理拓扑上建立无环路的逻辑拓扑。不属于活动的无环路拓扑的那一部分链路、端口和交换机并不真正参与数据帧的转发。当活动拓扑的一部分发生故障时,必须重新建立一个新的无环路的拓扑。这时,需要尽可能快地重新计算新的无环路拓扑或者使新的无环路拓扑收敛,以便减少终端用户无法访问网络资源的时间。

9.1.2　STP 算法

当网络稳定时,网络也已收敛,此时,每一个网络都有一棵生成树,见图 9.1。因此,对于每一个交换网络,都有一个根网桥,每个非根网桥都有一个根端口,每个网段都有一个指定端口,不使用非指定端口。根端口和指定端口用于转发(Forwarding)数据流量,非指定端

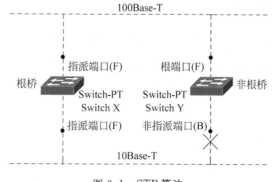

图 9.1　STP 算法

口将丢弃数据流量,它们也称为阻塞(Blocking)端口。网桥的根端口是最接近根网桥的端口,每个非根网桥都必须选出一个根端口。

为使采用 STP 的网络最初收敛为一个逻辑上无环路的网络拓扑,可以通过以下三步实现:

(1) 选举一个根网桥。这个协议有一个选举根网桥的过程。在一个给定网络中,只有一台网桥可以作为根网桥。根网桥的所有端口都是指定端口,指定端口通常处于转发状态。当端口处于转发状态时,它可以发送和接收流量。

(2) 在非根网桥上选择根端口。STP 在每个非根网桥上建立一个根端口。这个根端口是从非根网桥到根网桥的最低成本路径。根端口一般处在转发状态。生成树的路径成本是从根网桥到该非根网桥路径上所有成本的一个累加值,这些成本是基于带宽计算出来的。

(3) 为每个网段选举指定端口。在每个网段上,STP 只选举一个指定端口。将网桥上到达根网桥有最低成本的端口选为指定端口。指定端口一般处于转发状态,为该网段转发流量。

9.1.3 网桥协议数据单元

STP 需要网络设备互相交换消息来检测桥接环路,交换机发送的用于构建无环路拓扑的消息称为网桥协议数据单元(Bridge Protocol Data Unit,BPDU)。端口会不断收到BPDU,以保证在活动路径或设备发送故障的时候,仍然可以计算出一棵新的生成树。

BPDU 包含足够的信息,因此所有交换机利用这些信息可以完成以下的工作:

(1) 选择一台单独的交换机作为生成树的根。

(2) 计算它自身到根交换机的最短路径。

(3) 对于每一个 LAN 网段,指定一台交换机作为最接近根的交换机,称它为指定交换机。指定交换机处理所有从 LAN 到根交换机的通信。

(4) 每个非根交换机选择自身一个端口作为根端口,它是到根交换机路径最短的接口。

(5) 在每个网段上选择属于生成树一部分的端口作为指定端口。非指定端口将被阻塞。

生成树构造一个无环路拓扑时,它总是使用相同的 4 步来判定:

(1) 最低的根网桥 ID(Bridge ID,BID)。

(2) 到根网桥的最低路径成本。

(3) 最低的发送网桥 ID。

(4) 最低的端口 ID。

网桥使用这 4 步来保存各个端口中接收到的最佳的 BPDU 的一个副本。其中,根网桥 ID 的作用是通告网络中的根网桥;到根网桥路径成本表示发送此 BPDU 的交换机根端口到根桥的开销;发送网桥 ID 是发送此 BPDU 的交换机的 ID;端口 ID 表示此 BPDU 是从哪个端口发出的。

9.1.4 端口的状态

生成树协议的端口状态有如下 4 种。

(1) 阻塞(Blocking)。阻塞状态并不是物理地使端口关闭,而是逻辑地使端口处于不收发数据帧的状态。但是，BPDU 数据帧即使是阻塞状态的端口也是允许通过的。交换机依靠 BPDU 互相学习信息，阻塞的端口必须允许这样的数据帧通过，所以阻塞的端口实际上还是激活的。当网络里的交换机刚刚启动的时候，所有的端口都处于阻塞的状态，这种状态要维持 20s 的时间。这是为了防止在启动过程中产生交换环路。

（2）监听（Listening）。阻塞状态后，端口的状态变为监听状态，交换机开始互相学习 BPDU 里的信息。这个状态要维持 15s，以便交换机可以学习到网络里所有其他交换机的信息。在这个状态中，交换机不能转发数据帧，也不能进行 MAC 地址与端口的映射，MAC 地址的学习是不可能的。

（3）学习（Learning）。监听状态后，端口的状态变为学习状态。在这个状态中，交换机对学习到的其他交换机的信息进行处理，开始计算生成树协议。在这个状态中，已经开始允许交换机学习 MAC 地址，进行 MAC 地址与端口的映射，但是交换机还是不能转发数据帧。这个状态也要维持 15s，以便网络中所有的交换机都可以计算完毕。

（4）转发（Forwarding）。当学习状态结束后，交换机已经完成了生成树协议的计算，所有应该进入转发状态的端口转变为转发状态，应该进入阻塞状态的端口进入阻塞状态，交换机开始正常工作。

综上所述，阻塞状态和转发状态是生成树协议的一般状态，监听状态和学习状态是生成树协议的过渡状态。

当出现网络故障时，发现该故障的交换机会向根交换机发送 BPDU，根交换机会向其他交换机发出 BPDU 通告该故障，所有收到该 BPDU 的交换机会把自己的端口全部置为阻塞状态，然后重复上面叙述的过程，直到收敛。

9.1.5 选举根网桥

确定根网桥的算法，是比较交换机之间的 BID，BID 为优先级加 MAC 地址所得来的值。Cisco 的交换机的优先级可以是 0～65 535 范围里的值。但是由于 Cisco 的交换机默认的优先级是 32 768，如果不使用命令改变优先级，所有 Cisco 的交换机的优先级都是一样的。结果，在确定根网桥时，往往是比较网桥的 MAC 地址，MAC 地址最小的交换机就成为根网桥。如果想要人为让某台交换机成为根网桥，那么需要改变交换机的优先级。

9.1.6 选举根端口

每一台非根网桥的交换机，都有一个端口称为根端口。根端口是该交换机上到达根网桥路径成本最小的端口，该端口不能被阻塞。交换机上的每个端口都有端口成本，它的大小根据端口所连接的介质速度不同而不同。那么，端口上的路径成本就是到达某个目的设备的路径上一系列端口开销的和。

9.1.7 选举指定端口

桥接网络中的每个网段都有一个指定端口。该端口起到单独的桥接端口的作用，也就是负责发送和接收在该网段和根网桥之间的流量。这样设置的原因是：如果只有一个端口来处理每一个连接的流量，所有的环路都可以打破。包含指定端口的网桥称为该网段的指定网桥。选举根端口的同时，基于到根网桥的根路径成本的累计值的指定端口选择过程也在进行。

9.2 实验 1 生成树

9.2.1 实验拓扑

交换机 Switch0、Switch1 和 Switch2 通过 100Mb/s 链路彼此连接,未作任何配置,见图 9.2。

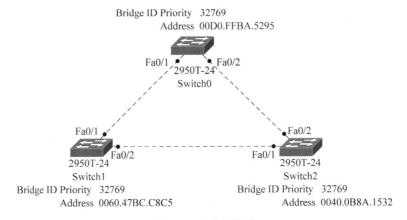

图 9.2 生成树协议

9.2.2 实验内容

(1) 默认配置下,查看生成树信息。
(2) 配置交换机生成树角色,查看生成树信息。
(3) 修改交换机默认优先级,查看生成树信息。
(4) 修改交换机端口默认路径成本,查看生成树信息。
(5) 修改交换机端口默认优先级,查看生成树信息。
(6) 修改交换机链路状态,查看生成树信息。

9.2.3 实验步骤

(1) 查看交换机 Switch0、Switch1 和 Switch2 上的生成树信息。

```
Switch0:
Switch0 > enable
Switch0 # show sp
Switch0 # show spanning - tree
VLAN0001
  Spanning tree enabled protocol ieee
  Root ID    Priority    32769
             Address     0040.0B8A.1532
             Cost        19
             Port        2(FastEthernet0/2)
```

```
                    Hello Time      2 sec Max Age 20 sec Forward Delay 15 sec

    Bridge ID       Priority        32769     (priority 32768 sys-id-ext 1)
                    Address         00D0.FFBA.5295
                    Hello Time      2 sec    Max Age 20 sec Forward Delay 15 sec
                    Aging Time      20

Interface       Role Sts Cost       Prio.Nbr    Type
--------        ------------        ---------   ------
Fa0/2           Root FWD 19         128.2       P2p
Fa0/1           Altn BLK 19         128.1       P2p
```

```
Switch1:
Switch1 > enable
Switch1 # show spanning-tree
VLAN0001
  Spanning tree enabled protocol ieee
  Root ID       Priority        32769
                Address         0040.0B8A.1532
                Cost            19
                Port            2(FastEthernet0/2)
                Hello Time      2 sec Max Age 20 sec Forward Delay 15 sec

  Bridge ID     Priority        32769 (priority 32768 sys-id-ext 1)
                Address         0060.47BC.C8C5
                Hello Time      2 sec Max Age 20 sec Forward Delay 15 sec
                Aging Time      20

Interface       Role Sts Cost       Prio.Nbr    Type
--------        ------------        ---------   ------
Fa0/1           Desg FWD 19         128.1       P2p
Fa0/2           Root FWD 19         128.2       P2p
```

```
Switch2:
Switch2 > enable
Switch2 # show span
Switch2 # show spanning-tree
VLAN0001
  Spanning tree enabled protocol ieee
  Root ID       Priority        32769
                Address         0040.0B8A.1532
                This bridge is the root
                Hello Time      2 sec Max Age 20 sec Forward Delay 15 sec

  Bridge ID     Priority        32769 (priority 32768 sys-id-ext 1)
                Address         0040.0B8A.1532
                Hello Time      2 sec Max Age 20 sec Forward Delay 15 sec
                Aging Time      20
```

```
Interface       Role Sts Cost    Prio.Nbr    Type
---------       -------------    --------    ----
Fa0/1           Desg FWD 19      128.1       P2p
Fa0/2           Desg FWD 19      128.2       P2p
```

【提示】 STP 是基于 VLAN 的,可以看到 STP 在 VLAN0001 上运行,协议版本是 IEEE 802.1D。

【提示】 网桥 BID 由优先级和 MAC 地址两部分组成,BID 最小的网桥称为根网桥,由于所有交换机均为默认配置,其优先级皆为 32 769,交换机 Switch2 的 MAC 地址最小,为 0040.0B8A.1532,可以看到根网桥的 BID 与交换机 Switch2 的 BID 一致,即交换机 Switch2 被选为根网桥。

【提示】 非根网桥 Switch0 和 Switch1 的各自所有端口中,FastEthernet0/2 端口只需通过一条 100Mb/s 的链接就可到达根网桥,路径成本 19 为最低(FastEthernet0/1 端口需要通过两条 100Mb/s 的链接就可到达根网桥,路径成为 19+19),即非根网桥 Switch0 和 Switch1 的 FastEthernet0/2 端口被选为各自的根端口(Root),处于转发状态(FWD)。

【提示】 根网桥的所有端口都是指定端口(Desg),处于转发状态(FWD),对于非根网桥 Switch0 和 Switch1 之间的链路,Switch0 和 Switch1 具有相同的根网桥 BID,各自需要通过一条 100Mb/s 的链接到达根网桥,具有相同的路径成本为 19,但 Switch1 的 BID 比 Switch0 的 BID 小,Switch1 的 FastEthernet0/1 端口被选为该链路的指定端口(Desg),处于转发状态(FWD),而 Switch0 的 FastEthernet0/1 端口被选为根端口 FastEthernet0/2 的替换端口(Altn),处于阻塞状态(BLK)。

【提示】 最终确定的无环路径如图 9.3 所示。

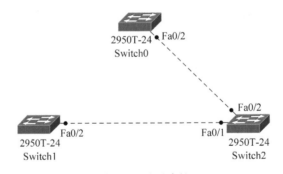

图 9.3 无环路径

(2) 配置交换机 Switch0 生成树角色为根网桥,查看生成树信息。

```
Switch0:
Switch0#configure terminal
Enter configuration commands, one per line. End with CNTL/Z.
Switch0(config)#spanning-tree vlan 1 root primary
Switch0(config)#end
Switch0#
%SYS-5-CONFIG_I: Configured from console by console
```

```
Switch0#show spanning-tree
VLAN0001
  Spanning tree enabled protocol ieee
  Root ID    Priority       24577
             Address        00D0.FFBA.5295
             This bridge is the root
             Hello Time 2 sec Max Age 20 sec Forward Delay 15 sec

  Bridge ID  Priority       24577 (priority 24576 sys-id-ext 1)
             Address        00D0.FFBA.5295
             Hello Time     2 sec Max Age 20 sec Forward Delay 15 sec
             Aging Time     20

Interface    Role Sts Cost    Prio.Nbr    Type
--------     ------------     --------    ------
Fa0/1        Desg FWD 19      128.1       P2p
Fa0/2        Desg FWD 19      128.2       P2p
```

```
Switch1:
Switch1#show spanning-tree
VLAN0001
  Spanning tree enabled protocol ieee
  Root ID    Priority       24577
             Address        00D0.FFBA.5295
             Cost           19
             Port           1(FastEthernet0/1)
             Hello Time     2 sec Max Age 20 sec Forward Delay 15 sec

  Bridge ID  Priority       32769 (priority 32768 sys-id-ext 1)
             Address        0060.47BC.C8C5
             Hello Time     2 sec Max Age 20 sec Forward Delay 15 sec
             Aging Time     20

Interface    Role Sts Cost    Prio.Nbr    Type
--------     ------------     --------    ------
Fa0/1        Root FWD 19      128.1       P2p
Fa0/2        Altn BLK 19      128.2       P2p
```

```
Switch2:
Switch2#show spanning-tree
VLAN0001
  Spanning tree enabled protocol ieee
  Root ID    Priority       24577
             Address        00D0.FFBA.5295
             Cost           19
             Port           2(FastEthernet0/2)
             Hello Time     2 sec Max Age 20 sec Forward Delay 15 sec
```

```
Bridge ID    Priority     32769 (priority 32768 sys-id-ext 1)
             Address      0040.0B8A.1532
             Hello Time   2 sec  Max Age 20 sec  Forward Delay 15 sec
             Aging Time   20

Interface    Role Sts Cost    Prio.Nbr    Type
---------    -------------    --------    ------
Fa0/1        Desg FWD 19      128.1       P2p
Fa0/2        Root FWD 19      128.2       P2p
```

【提示】 可以通过命令，自动降低交换机当前 BID 的优先级（降低数值为 4096 的倍数，直至新的 BID 比所有其他交换机的 BID 都低），强制该交换机成为根网桥。

【提示】 最终确定的无环路径如图 9.4 所示。

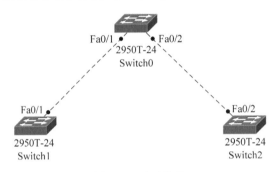

图 9.4　无环路径

（3）恢复所有交换机默认配置，配置交换机 Switch0 生成树角色为辅助根网桥，查看生成树信息。

```
Switch0:
Switch0#config terminal
Enter configuration commands, one per line. End with CNTL/Z.
Switch0(config)#spanning-tree vlan 1 root secondary
Switch0(config)#end
Switch0#
%SYS-5-CONFIG_I: Configured from console by console
Switch0#show spanning-tree
VLAN0001
  Spanning tree enabled protocol ieee
  Root ID    Priority    28673
             Address     00D0.FFBA.5295
             This bridge is the root
             Hello Time  2 sec  Max Age 20 sec  Forward Delay 15 sec

  Bridge ID  Priority    28673 (priority 28672 sys-id-ext 1)
             Address     00D0.FFBA.5295
             Hello Time  2 sec  Max Age 20 sec  Forward Delay 15 sec
             Aging Time  20
```

```
Interface        Role Sts Cost     Prio.Nbr       Type
----------       ------------      ----------     ------
Fa0/1            Desg LSN 19       128.1          P2p
Fa0/2            Desg FWD 19       128.2          P2p
```

Switch1:
```
    Switch1#show spanning-tree
VLAN0001
  Spanning tree enabled protocol ieee
  Root ID     Priority       28673
              Address        00D0.FFBA.5295
              Cost           19
              Port           1(FastEthernet0/1)
              Hello Time     2 sec Max Age 20 sec Forward Delay 15 sec

  Bridge ID   Priority       32769 (priority 32768 sys-id-ext 1)
              Address        0060.47BC.C8C5
              Hello Time     2 sec Max Age 20 sec Forward Delay 15 sec
              Aging Time     20

Interface        Role Sts Cost     Prio.Nbr       Type
----------       ------------      ----------     ------
Fa0/1            Root FWD 19       128.1          P2p
Fa0/2            Altn BLK 19       128.2          P2p
```

Switch2:
```
    Switch2#show spanning-tree
VLAN0001
  Spanning tree enabled protocol ieee
  Root ID     Priority       28673
              Address        00D0.FFBA.5295
              Cost           19
              Port           2(FastEthernet0/2)
              Hello Time     2 sec Max Age 20 sec Forward Delay 15 sec

  Bridge ID   Priority       32769 (priority 32768 sys-id-ext 1)
              Address        0040.0B8A.1532
              Hello Time     2 sec Max Age 20 sec Forward Delay 15 sec
              Aging Time     20

Interface        Role Sts Cost     Prio.Nbr       Type
----------       ------------      ----------     ------
Fa0/1            Desg FWD 19       128.1          P2p
Fa0/2            Root FWD 19       128.2          P2p
```

【提示】 可以通过命令，自动降低交换机和根网桥当前 BID 的优先级（降低数值为 4096 的倍数，直至交换机新的 BID 比除了根网桥之外所有其他交换机的 BID 都低），强制该交换机成为辅助根网桥，一旦根网桥失效，能够保证该交换机成为根网桥。

【提示】 配置辅助根网桥，不会影响当前的无环路径。

（4）恢复所有交换机默认配置，修改交换机 Switch1 优先级为 28 672（比当前根网桥 Switch2 优先级低 4096），查看生成树信息。

```
Switch1:
Switch1#configure terminal
Enter configuration commands, one per line. End with CNTL/Z.
Switch1(config)#spanning-tree vlan 1 priority 28672
Switch1(config)#end
Switch1#
%SYS-5-CONFIG_I: Configured from console by console

Switch1#show spanning-tree
VLAN0001
  Spanning tree enabled protocol ieee
  Root ID    Priority     28673
             Address      0060.47BC.C8C5
             This bridge is the root
             Hello Time   2 sec  Max Age 20 sec  Forward Delay 15 sec

  Bridge ID  Priority     28673 (priority 28672 sys-id-ext 1)
             Address      0060.47BC.C8C5
             Hello Time   2 sec  Max Age 20 sec  Forward Delay 15 sec
             Aging Time   20

Interface    Role Sts Cost      Prio.Nbr   Type
--------     ------------       ----------  ------
Fa0/1        Desg FWD 19        128.1      P2p
Fa0/2        Altn BLK 19        128.2      P2p


Switch0:
Switch0#show spanning-tree
VLAN0001
  Spanning tree enabled protocol ieee
  Root ID    Priority     28673
             Address      0060.47BC.C8C5
             Cost         19
             Port         1(FastEthernet0/1)
             Hello Time   2 sec  Max Age 20 sec  Forward Delay 15 sec

  Bridge ID  Priority     32769 (priority 32768 sys-id-ext 1)
             Address      00D0.FFBA.5295
             Hello Time   2 sec  Max Age 20 sec  Forward Delay 15 sec
             Aging Time   20

Interface    Role Sts Cost      Prio.Nbr   Type
--------     ------------       ----------  ------
Fa0/1        Root FWD 19        128.1      P2p
Fa0/2        Altn BLK 19        128.2      P2p
```

```
Switch2:
Switch2#show spanning-tree
VLAN0001
  Spanning tree enabled protocol ieee
  Root ID    Priority        28673
             Address         0060.47BC.C8C5
             Cost            19
             Port            1(FastEthernet0/1)
             Hello Time      2 sec Max Age 20 sec Forward Delay 15 sec

  Bridge ID  Priority        32769 (priority 32768 sys-id-ext 1)
             Address         0040.0B8A.1532
             Hello Time      2 sec Max Age 20 sec Forward Delay 15 sec
             Aging Time      20

Interface    Role Sts Cost   Prio.Nbr   Type
----------   ------------    ---------  ------
Fa0/1        Root FWD 19     128.1      P2p
Fa0/2        Desg FWD 19     128.2      P2p
```

【提示】 指定的交换机优先级必须为4096的倍数。

【提示】 最终确定的无环路径如图9.5所示。

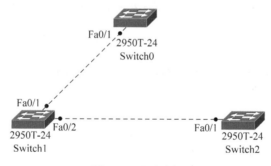

图9.5 无环路径

(5) 恢复所有交换机默认配置,修改交换机 Switch0 的 FastEthernet0/2 端口路径成本为100(相当于10Mb/s端口带宽),查看生成树信息。

```
Switch0:
Switch0#configure terminal
Enter configuration commands, one per line. End with CNTL/Z.
Switch0(config)#interface fastEthernet 0/2
Switch0(config-if)#spanning-tree vlan 1 cost 100
Switch0(config-if)#end
Switch0#show spanning-tree
VLAN0001
  Spanning tree enabled protocol ieee
  Root ID    Priority        32769
             Address         0040.0B8A.1532
```

```
              Cost          38
              Port          1(FastEthernet0/1)
              Hello Time    2 sec Max Age 20 sec Forward Delay 15 sec

   Bridge ID  Priority      32769 (priority 32768 sys-id-ext 1)
              Address       00D0.FFBA.5295
              Hello Time    2 sec Max Age 20 sec Forward Delay 15 sec
              Aging Time    20

Interface    Role Sts Cost    Prio.Nbr    Type
---------    ------------    ----------   ------
Fa0/1        Root FWD 19      128.1       P2p
Fa0/2        Altn BLK 100     128.2       P2p3.3
```

Switch1:
```
Switch1# show spanning-tree
VLAN0001
  Spanning tree enabled protocol ieee
  Root ID    Priority      32769
             Address       0040.0B8A.1532
             Cost          19
             Port          2(FastEthernet0/2)
             Hello Time    2 sec Max Age 20 sec   Forward Delay 15 sec

Bridge ID Priority          32769  (priority 32768 sys-id-ext 1)
             Address        0060.47BC.C8C5
             Hello Time     2 sec  Max Age 20 sec  Forward Delay 15 sec
             Aging Time     20

Interface    Role Sts Cost    Prio.Nbr    Type
---------    ------------    ----------   ------
Fa0/1        Desg FWD 19      128.1       P2p
Fa0/2        Root FWD 19      128.2       P2p
```

Switch2:
```
Switch2# show spanning-tree
VLAN0001
  Spanning tree enabled protocol ieee
  Root ID    Priority      32769
             Address       0040.0B8A.1532
             This bridge is the root
             Hello Time    2 sec Max Age 20 sec Forward Delay 15 sec

   Bridge ID  Priority     32769 (priority 32768 sys-id-ext 1)
              Address      0040.0B8A.1532
              Hello Time   2 sec Max Age 20 sec Forward Delay 15 sec
              Aging Time   20
```

```
Interface        Role Sts Cost    Prio.Nbr    Type
---------        ------------    ---------    ----
Fa0/1            Desg FWD 19      128.1       P2p
Fa0/2            Desg FWD 19      128.2       P2p
```

【提示】 对于非根网桥 Switch0 的 FastEthernet0/1 到根网桥需要经过两个 100Mb/s 链路，路径成本为 19+19=38，而 FastEthernet0/2 到根网桥路径成本根据配置为 100，所以 FastEthernet0/1 被选为根端口（Root），处于转发状态（FWD），FastEthernet0/2 被选为根端口的替换端口（Altn），处于阻塞状态（BLK）。

【提示】 最终确定的无环路径如图 9.6 所示。

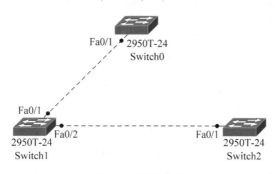

图 9.6　无环路径

（6）恢复所有交换机默认配置，更改实验拓扑如图 9.7 所示，查看交换机 Switch1 生成树信息。

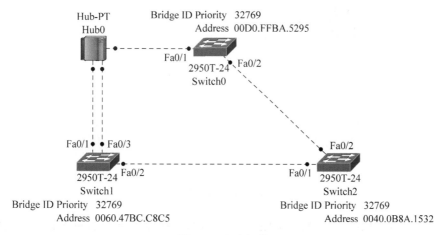

图 9.7　实验拓扑

```
Switch1:
Switch1#show spanning-tree
VLAN0001
  Spanning tree enabled protocol ieee
  Root ID     Priority      32769
              Address       0040.0B8A.1532
```

```
            Cost            19
            Port            2(FastEthernet0/2)
            Hello Time 2 sec Max Age 20 sec Forward Delay 15 sec

Bridge ID   Priority        32769 (priority 32768 sys-id-ext 1)
            Address         0060.47BC.C8C5
            Hello Time 2 sec Max Age 20 sec Forward Delay 15 sec
            Aging Time 20

Interface    Role Sts Cost    Prio.Nbr    Type
--------     ------------    ----------   ------
Fa0/1        Desg FWD 19      128.1       Shr
Fa0/2        Root FWD 19      128.2       P2p
Fa0/3        Altn BLK 19      128.3       Shr
```

【提示】 对于 Switch1 和 Switch0 之间的链路,因为 Switch1 的 BID 小于 Switch0 的 BID,该链路的指定端口在 Switch1 上选择,又因为 Switch1 上的 FastEthernet0/1 端口的 PID(128.1)小于 FastEthernet0/3 端口的 PID(128.3),即两个端口具有相同的端口优先级,只能根据端口号进行 PID 的区分,最后 FastEthernet0/1 端口被选为该链路的指定端口 (Desg),处于转发状态(FWD),而 FastEthernet0/3 端口被选为根端口的替代端口(Altn),处于阻塞状态(BLK)。

【提示】 最终确定的无环路径如图 9.8 所示。

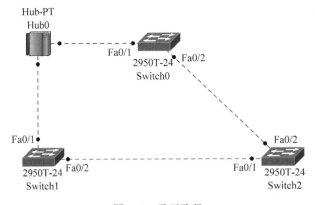

图 9.8 无环路径

(7) 修改交换机 Switch1 的 FastEthernet0/3 端口优先级为 112,再次查看生成树信息。

```
Switch1:
Switch1#configure terminal
Enter configuration commands, one per line. End with CNTL/Z.
Switch1(config)#interface fastEthernet 0/3
Switch1(config-if)#spanning-tree vlan 1 port-priority 112
Switch1(config-if)#end
Switch1#
```

```
% SYS-5-CONFIG_I: Configured from console by console

Switch1#show spanning-tree
VLAN0001
  Spanning tree enabled protocol ieee
  Root ID    Priority     32769
             Address      0040.0B8A.1532
             Cost         19
             Port         2(FastEthernet0/2)
             Hello Time   2 sec   Max Age 20 sec   Forward Delay 15 sec

  Bridge ID  Priority     32769 (priority 32768 sys-id-ext 1)
             Address      0060.47BC.C8C5
             Hello Time   2 sec   Max Age 20 sec   Forward Delay 15 sec
             Aging Time 20

Interface      Role Sts Cost     Prio.Nbr     Type
--------       ------------      ----------   ------
Fa0/1          Altn BLK 19       128.1        Shr
Fa0/2          Root FWD 19       128.2        P2p
Fa0/3          Desg FWD 19       112.3        Shr
```

【提示】 由于更改了 FastEthernet0/3 端口的优先级,其端口优先级 112 小于 FastEthernet0/1 端口 128,尽管其端口号 3 大于 FastEthernet0/1 端口号 1,但是其 PID 小于 FastEthernet0/1 的 PID,所以 FastEthernet0/3 端口被选为该链路的指定端口(Desg),处于转发状态(FWD),而 FastEthernet0/1 端口被选为根端口的替换端口(Altn),处于阻塞状态(BLK)。

【提示】 最终确定的无环路径如图 9.9 所示。

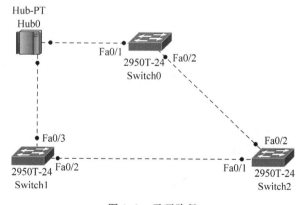

图 9.9 无环路径

(8) 实验拓扑如图 9.10 所示,所有交换机均为默认配置,当前确定的无环路径如图 9.11 所示,通过关闭交换机 Switch3 的端口改变交换机 Switch3 和交换机 Switch4 之间的链路状态,再次确定的无环路径如图 9.12 所示。

【提示】 生成树协议在链路状态发生变化时,会自动重新进行计算,确认新的无环路径。

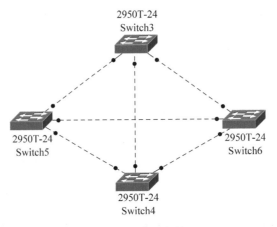

图 9.10 实验拓扑

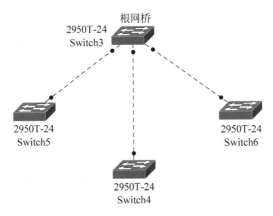

图 9.11 无环路径

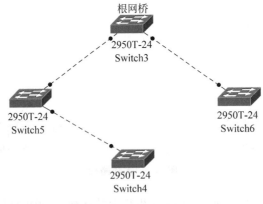

图 9.12 无环路径

第 10 章　虚拟网间路由

10.1　基础知识

10.1.1　一般路由器实现 VLAN 间路由

每一个 VLAN 是一个广播域，不同 VLAN 里的主机如果不通过路由器，是不能通信的。两台交换机之间可以通过干道的连接，达到两台交换机所连接的相同 VLAN 里的主机互相通信的目的。但是，两台交换机上不同 VLAN 的主机之间还是不能通信。要想让两台属于不同 VLAN 的主机之间能够通信，必须使用路由器为 VLAN 之间作路由，如图 10.1 所示。

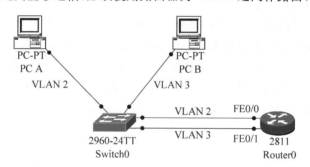

图 10.1　不同 VLAN 的主机通过路由器互相通信

在图 10.1 中，交换机使用两个分别属于 VLAN 2 和 VLAN 3 的端口连接路由器的两个以太网端口 FE0/0 和 FE0/1。在路由器的 FE0/0 接口上配置与 VLAN 2 里的主机同网段的 IP 地址，在路由器的 FE0/1 接口上配置与 VLAN 3 里的主机同网段的 IP 地址。在主机 A 上设置默认网关为路由器的 FE0/0 接口地址，在主机 B 上设置默认网关为路由器的 FE0/1 接口地址，则主机 A 与主机 B 可以实现互相通信，因为交换机把来自于主机 A 去往主机 B 的数据帧通过 VLAN 2 的端口发送给路由器的 FE0/0 端口。路由器是工作在三层上的设备，它根本不管数据帧是从哪个 VLAN 来的，它只是按照数据包 IP 包头里封装的目的 IP 地址为数据包作路由。由于数据包头里的目的 IP 地址（即主机 B 的 IP 地址）位于路由器 FE0/1 接口所连接的网段上，路由器把该数据包由 FE0/1 接口发出，该数据包从交换机的属于 VLAN 3 的端口到达交换机，交换机把该数据包转发给主机 B。主机 B 与主机 A 之间的通信是上述过程的逆过程。所以，VLAN 间的路由实际上就是子网间的路由。

10.1.2　单臂路由器实现 VLAN 间路由

但是，如果网络中有很多 VLAN，需要很多交换机和路由器之间的连接，一般来说，大多数路由器没有很多个接口。所以，一般用一条以太线或者光纤以干道的方式连接交换机和路由器，因为在干道上可以传输多个VLAN的数据。要想使用干道连接交换机和路由器，路由器和交换机两端的接口都必须是100Mb/s以上的以太接口或光纤接口。

如图10.2所示的设计就可以为多个VLAN作路由。

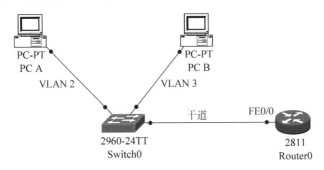

图10.2 干道连接交换机和路由器

把每个VLAN与路由器有一条线路连接的、为VLAN间提供路由的方式称为物理连接方式；把使用干道在路由器和交换机之间传递多个VLAN信息的、为VLAN间提供路由的方式称为逻辑连接方式。

如果只使用一条线路作为干道连接路由器和交换机，路由器上只有一个物理接口和交换机连接，在该物理接口上有效的IP地址只能配置一个。可是要求每一个VLAN对应一个子网。在干道上要求有多条逻辑的线路和路由器连接，即干道上允许多少个VLAN通过，就应该有多少条逻辑的线路。每条VLAN的逻辑线路，都要求和路由器上IP地址属于该VLAN所在子网的接口连接。如图10.3所示方法解决了多个VLAN和一个物理接口对应的问题。

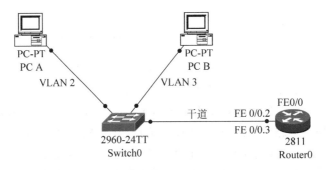

图10.3 在物理接口上划分子接口

在图10.3中，可以看到路由器的物理接口FE0/0被逻辑地划分成了两个子接口FE0/0.2和FE0/0.3，每一个子接口对应一个VLAN的子网。从而使干道上的每一条逻辑线路都有了一个路由器的接口与之对应，只不过这些接口都是虚拟的逻辑接口。但是路由器视这些接口为正常接口，其功能与一般物理接口一样。

在路由器的一个物理接口上，可以存在多个逻辑的子接口。可以在一个物理接口上配置与VLAN数量相当的子接口，使每一个VLAN的子网连接在一个子接口上。为每一个子接口分配其对应VLAN的子网中的IP地址，而不在物理接口上分配IP地址。

在图 10.3 中,把路由器上连接干道的物理接口分成了两个子接口,以对应网络中的两个 VLAN。其中,子接口 FE0/0.2 对应 VLAN 2,另一个子接口 FE0/0.3 对应 VLAN 3。如果 VLAN 2 的网段是 192.168.2.0,VLAN 3 的网段是 192.168.3.0,那么子接口 FE0/0.2 可被配置 192.168.2.0 网段的地址 192.168.2.1,而子接口 FE0/0.3 可被配置 192.168.3.0 网段的地址 192.168.3.1。

这样,路由器会认为自己的两个子接口直接连接了两个网段 192.168.2.0 和 192.168.3.0。路由器上的接口所直接连接的网段会直接进入路由表。所以,经过上述配置,路由器已经具备了为两个 VLAN 里的数据包进行路由的条件了。

但是,为了让 VLAN 之间能够通信,还要在 VLAN 里的主机上配置默认网关,以使 VLAN 里的主机在向其他 VLAN 的主机发送数据时,要将数据包发送到路由器上来。主机的默认网关地址,就是路由器连接该主机所在 VLAN 的子接口地址。

10.1.3 三层交换机实现 VLAN 间路由

单臂路由器实现 VLAN 间的路由时转发速度较慢,实际上,在局域网内部多采用三层交换。三层交换机通常采用硬件来实现,其路由数据包的速率是普通路由器的几十倍。从使用者的角度,可以把三层交换机看成二层交换机和路由器的组合。如图 10.4 所示,这个虚拟的路由器和每个 VLAN 都有一个接口进行连接,不过这些接口的名称是 VLAN1 或者 VLAN2。思科早些年采用的基于 NetFlow 的三层交换技术,现在思科主要采用 CEF 技术。在 CEF 技术中,交换机利用路由表形成转发信息库(FIB),FIB 和路由表是同步的,关键的是 FIB 的查询是硬件化的,其查询速度很快。除了 FIB 外,还有邻接表(Adjacency Table),该表和 ARP 表类似,主要放置了第二层的封装信息。FIB 和邻接表都是在数据转发之前就已经建立好了,这样一有数据要转发,交换机就能直接利用它们进行数据转发和封装,不需要查询路由表和发送 ARP 请求,所以 VLAN 间的路由速度大大提高。

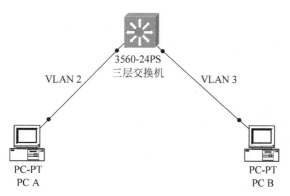

图 10.4 三层交换机实现 VLAN 间路由

10.2 实验 1 一般路由器实现 VLAN 间路由

10.2.1 实验拓扑

交换机 Switch 分别与两台不同 VLAN 中的 PC 连接,同时为了实现 VLAN 间通信与路由器 Router0 两个不同的物理接口连接,见图 10.5。

设备名	IP 地址	网关
PCA	192.168.2.1	192.168.2.2
PCB	192.168.3.1	192.168.3.2
Router:Fa0/0	192.168.3.2	
Router:Fa0/1	192.168.2.2	

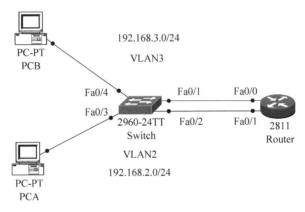

图 10.5 实验拓扑图

10.2.2 实验内容

(1) 不经过路由器 VLAN 间通信测试。
(2) 经过路由器不同物理接口 VLAN 间通信配置。

10.2.3 实验步骤

(1) 在交换机 Switch0 上定义 VLAN2 和 VLAN3,并将 PCA 和 PCB 分配到 VLAN2 和 VLAN3,路由器物理接口属于关闭状态。

```
Switch>
Switch> enable
Switch# configure terminal
Enter configuration commands, one per line. End with CNTL/Z.
Switch(config)# vlan 2
Switch(config-vlan)# exit
Switch(config)# vlan 3
Switch(config-vlan)# exit
Switch(config)# interface fastEthernet 0/3
Switch(config-if)# switchport mode access
Switch(config-if)# switchport access vlan 2
Switch(config-if)# exit
Switch(config)# int fastEthernet 0/4
Switch(config-if)# switchport mode access
Switch(config-if)# switchport access vlan 3
Switch(config-if)# end
Switch#
```

（2）PCA 与 PCB 进行连通性测试。

```
   PCA:
PC>ping 192.168.3.1

Pinging 192.168.3.1 with 32 bytes of data:

Request timed out.
Request timed out.
Request timed out.
Request timed out.

Ping statistics for 192.168.3.1:
    Packets: Sent = 4, Received = 0, Lost = 4 (100% loss),
```

【提示】 不同 VLAN 间的通信不能直接经过交换机进行通信，即使在同一条交换机也不行。

（3）打开路由器的物理接口，并分配到 VLAN2 和 VLAN3。

```
Switch#configure terminal
Enter configuration commands, one per line. End with CNTL/Z.
Switch(config)#interface fastEthernet 0/1
Switch(config-if)#switchport mode access
Switch(config-if)#switchport access vlan 3
Switch(config-if)#exit
Switch(config)#interface fastEthernet 0/2
Switch(config-if)#switchport mode access
Switch(config-if)#switchport access vlan 2
Switch(config-if)#end
Switch#
```

（4）PCA 与 PCB 再次进行连通性测试。

```
PC>ping 192.168.3.1

Pinging 192.168.3.1 with 32 bytes of data:

Reply from 192.168.3.1: bytes = 32 time = 16ms TTL = 127
Reply from 192.168.3.1: bytes = 32 time = 9ms TTL = 127
Reply from 192.168.3.1: bytes = 32 time = 16ms TTL = 127
Reply from 192.168.3.1: bytes = 32 time = 24ms TTL = 127

Ping statistics for 192.168.3.1:
    Packets: Sent = 4, Received = 4, Lost = 0 (0% loss),
Approximate round trip times in milli-seconds:
Minimum = 9ms, Maximum = 24ms, Average = 16ms
```

【提示】 为了能够保证各个 VLAN 间的数据能够到达路由器的物理接口，需要交换机与路由器连接的物理接口分配到指定的 VLAN 上，也可以设置为干道端口，传输所有

VLAN 数据。

对于路由器只要存在到各个 VLAN 的路由即可实现 VLAN 间通信,本例中路由器到各个 VLAN 的路由属于直连路由,不需要进行配置,如果非直连路由,可以通过配置静态路由和进行动态路由学习建立到各个 VLAN 的路由。

10.3 实验 2 单臂路由实现 VLAN 间路由

10.3.1 实验拓扑

交换机 Switch 分别与两台不同 VLAN 中的 PC 连接,同时为了实现 VLAN 间通信与路由器 Router0 一个物理接口连接,见图 10.6。

设备名	IP 地址	网关
PCA	192.168.2.1	192.168.2.2
PCB	192.168.3.1	192.168.3.2
Router:Fa0/0.2	192.168.3.2	
Router:Fa0/0.3	192.168.2.2	

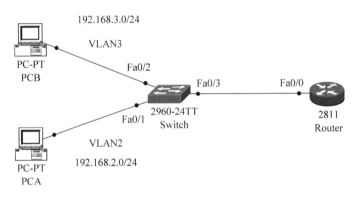

图 10.6 实验拓扑图

10.3.2 实验内容

经过路由器一个物理接口 VLAN 间通信配置。

10.3.3 实验步骤

(1) 在交换机 Switch 上定义 VLAN2 和 VLAN3,并将 PCA 和 PCB 分配到 VLAN2 和 VLAN3,配置 Fastethernet0/3 为干道端口。

```
Switch#configure terminal
Enter configuration commands, one per line. End with CNTL/Z.
Switch(config)#vlan 2
Switch(config-vlan)#exit
Switch(config)#vlan 3
```

```
Switch(config-vlan)# exit
Switch(config)# int
Switch(config)# interface fastEthernet 0/1
Switch(config-if)# switchport mode access
Switch(config-if)# switchport access vlan 2
Switch(config-if)# exit
Switch(config)# interface fastEthernet 0/2
Switch(config-if)# switchport mode access
Switch(config-if)# switchport access vlan 3
Switch(config-if)# exit
Switch(config)# interface fastEthernet 0/3
Switch(config-if)# switchport mode trunk
Switch(config-if)# end
Switch#
```

(2) 在路由器 Router 上定义子接口，配置指定 VLAN 封装。

```
Router# configure terminal
Enter configuration commands, one per line. End with CNTL/Z.
Router(config)# interface fastEthernet 0/0
Router(config-if)# no shutdown
Router(config-if)# exit
Router(config)# interface fastEthernet 0/0.2
Router(config-subif)# encapsulation dot1Q 2
Router(config-subif)# ip address 192.168.2.2 255.255.255.0
Router(config-subif)# exit
Router(config)# interface fastEthernet 0/0.3
Router(config-subif)# encapsulation dot1Q 3
Router(config-subif)# ip address 192.168.3.2 255.255.255.0
Router(config-subif)# exit
Router(config)# end
Router#
```

(3) PCA 与 PCB 进行连通性测试。

```
PC> ping 192.168.3.1

Pinging 192.168.3.1 with 32 bytes of data:

Reply from 192.168.3.1: bytes = 32 time = 16ms TTL = 127
Reply from 192.168.3.1: bytes = 32 time = 19ms TTL = 127
Reply from 192.168.3.1: bytes = 32 time = 15ms TTL = 127
Reply from 192.168.3.1: bytes = 32 time = 6ms TTL = 127

Ping statistics for 192.168.3.1:
    Packets: Sent = 4, Received = 4, Lost = 0 (0% loss),
Approximate round trip times in milli-seconds:
Minimum = 6ms, Maximum = 19ms, Average = 14ms
```

【提示】 为了在路由器使用子接口，首先要打开物理接口，在设置子接口 IP 地址前，需要先指定一下 dot1Q 封装，否则 IP 地址定义无效。

由于交换机与路由器使用一条链路进行连接，所以所有 VLAN 数据都需要经过该链路，要求交换机与该链路连接的端口必须为干道端口。

10.4 实验3 三层交换实现 VLAN 间路由

10.4.1 实验拓扑

三层交换机 Switch 分别与两台不同 VLAN 中的 PC 连接，见图 10.7。

设备名	IP 地址	网关
PCA	192.168.2.1	192.168.2.2
PCB	192.168.3.1	192.168.3.2
Switch：VLAN2	192.168.3.2	
Switch：VLAN3	192.168.2.2	

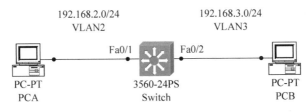

图 10.7 实验拓扑图

10.4.2 实验内容

经过三层交换机 VLAN 间通信配置。

10.4.3 实验步骤

(1) 在交换机 Switch 上定义 VLAN2 和 VLAN3，并将 PCA 和 PCB 分配到 VLAN2 和 VLAN3。

```
Switch > enable
Switch # configure terminal
Enter configuration commands, one per line. End with CNTL/Z.
Switch(config) # vlan 2
Switch(config - vlan) # exit
Switch(config) # vlan 3
Switch(config - vlan) # exit
Switch(config) # interface fastEthernet 0/1
Switch(config - if) # switchport mode access
Switch(config - if) # switchport access vlan 2
Switch(config - if) # exit
Switch(config) # interface fastEthernet 0/2
```

```
Switch(config-if)#switchport mode access
Switch(config-if)#switchport access vlan 3
Switch(config-if)#exit
%SYS-5-CONFIG_I: Configured from console by console
```

(2) 在交换机 Switch 上定义 VLAN2 和 VLAN3 的 IP 地址。

```
Switch(config)#interface vlan 2

%LINK-5-CHANGED: Interface Vlan2, changed state to up

%LINEPROTO-5-UPDOWN: Line protocol on Interface Vlan2, changed state to up
Switch(config-if)#ip address 192.168.2.2 255.255.255.0
Switch(config-if)#exit
Switch(config)#interface vlan 3
Switch(config-if)#

%LINK-5-CHANGED: Interface Vlan3, changed state to up

%LINEPROTO-5-UPDOWN: Line protocol on Interface Vlan3, changed state to up
Switch(config-if)#ip address 192.168.3.2 255.255.255.0
Switch(config-if)#exit
Switch(config)#
```

(3) 在交换机 Switch 上启用 IP 路由。

```
Switch(config)#ip routing
Switch(config)#end
Switch#
```

(4) PCA 与 PCB 进行连通性测试。

```
PC>ping 192.168.3.1

Pinging 192.168.3.1 with 32 bytes of data:

Reply from 192.168.3.1: bytes=32 time=16ms TTL=127
Reply from 192.168.3.1: bytes=32 time=19ms TTL=127
Reply from 192.168.3.1: bytes=32 time=15ms TTL=127
Reply from 192.168.3.1: bytes=32 time=6ms TTL=127

Ping statistics for 192.168.3.1:
    Packets: Sent = 4, Received = 4, Lost = 0 (0% loss),
Approximate round trip times in milli-seconds:
Minimum = 6ms, Maximum = 19ms, Average = 14ms
```

【提示】 三层交换机需要启动 IP 路由功能，并配置 VLAN2 和 VLAN3 的管理 IP 地址。

第11章　综合案例

11.1　案例需求

11.1.1　案例背景和目的

通过对辽宁科技大学等高校校园网分析，一般校园网主要需求包括如下几个方面。

(1) 网络设备需求：主要设备为路由器、三层交换机、百兆交换机。

(2) 网络可靠性和稳定行需求：网络为多出口、多核心结构，并配置有负载均衡与冗余备份。

(3) 网络安全需求：核心设备和接入设备支持 VLAN 划分，采用隔离技术保证网络带宽的充分利用，并有效地抑制病毒广播。

(4) 网络管理需求：具备网络拓扑发现、网络设备集中管理、性能监视和预警能力。

通过对辽宁科技大学校园网建设的实际建设方案分析，在最大程度保留实际建设方案需求背景基础上，结合前面各个章节介绍的路由技术和交换技术，有针对性地设计了一套适合学生进行综合实践的虚拟校园网设计案例，通过该案例可以将前面各个章节介绍的路由技术和交换技术综合在一起，锻炼学生从网络需求分析、网络拓扑设计到设备配置调试一整套的网络工程专业实践能力。

11.1.2　案例配置需求

(1) 有一个主校区，一个分校区和一个生活区，分别通过帧中继网络进行接入。

(2) 帧中继网络实现主校区、分校区和生活区快速通信。

(3) 配置动态和静态路由实现各个区域之间通信。

(4) 主校区内包括教学区、办公区、信息中心和宿舍区1。

教学区定义为一个 VLAN，通过高层交换机接入，主机地址静态分配。

办公区定义为一个 VLAN，通过高层交换机接入，主机地址静态分配，启动黏性安全 MAC 地址，设置最大安全 MAC 地址数和超限处理规则。

宿舍区1定义为一个 VLAN，通过高层交换机接入，主机地址静态分配。

教学区、办公区和独立宿舍区通过高层交换机实现 VLAN 间通信。

信息中心定义为一个物理 LAN，服务器地址静态分配，对校内和校外提供 Web 服务，只对校内提供 FTP 服务和 DNS 服务，接入互联网进行 NAT 转换。

(5) 分校区包括教学区和宿舍区2。

教学区定义为一个 VLAN，主机地址静态分配。

宿舍区2定义为一个 VLAN，主机地址动态分配。

教学区和宿舍区3通过单臂路由器实现VLAN间通信。

（6）生活区包括多个宿舍区，每个宿舍区定义为一个VLAN，动态地址分配，通过VTP进行VLAN管理，通过STP进行VLAN负载分担。

11.2 拓扑图设计

11.2.1 拓扑图

拓扑图如图11.1所示。

11.2.2 IP地址规划

IP地址规划如表11.1和表11.2所示。

表11.1 IP地址规划一

区域名	子区域名	分配方式	地址范围
主校区	信息中心	静态	192.168.5.0/24
	教学区1	静态	192.168.7.0/24
	办公区1	静态	192.168.8.0/24
	宿舍区1	静态	192.168.9.0/24
分校区	教学区2	动态	192.168.11.0/24
	宿舍区2	动态	192.168.12.0/24
生活区	宿舍区3	动态	192.168.14.0/24
	宿舍区4	动态	192.168.15.0/24
	宿舍区5	动态	192.168.16.0/24
	宿舍区6	动态	192.168.17.0/24

表11.2 IP地址规划二

区域名	设备名称	接口	IP地址	功能
信息中心	InforCenterSwitch			服务器和管理端接入
	WWW	FastEthernet	192.168.5.2/24	WWW服务器
	FTP	FastEthernet	192.168.5.3/24	FTP服务器
	DNS	FastEthernet	192.168.5.4/24	DNS服务器
	Manager	FastEthernet	192.168.5.5/24	管理端
	RouterA	FastEthernet0/0	192.168.2.1/24	静态路由；
		FastEthernet0/1	192.168.5.1/24	动态路由；
		FastEthernet1/0	192.168.6.1/24	外网接入；
		FastEthernet1/1	59.73.73.1/24	访问控制；NAT转换
	Internet	FastEthernet0/0	59.73.73.2/24	Internet接入
	MainAreaSwitch			区域接入；VLAN间路由；端口安全
教学区1	TeachArea1Switch			终端接入

续表

区域名	设备名称	接口	IP 地址	功 能
办公区 1	OfficeArea1Switch			终端接入
宿舍区 1	DormArea1Switch			终端接入
	RouterB	FastEthernet0/0	192.168.3.2	动态路由；
		FastEthernet0/1.1	192.168.11.1	DHCP 分配；
		FastEthernet0/1.2	192.168.12.1	VLAN 间路由
	BranchAreaSwitch			区域接入； 端口安全
教学区 2	TeachArea2Switch			终端接入
宿舍区 2	DormArea2Switch			终端接入
	RouterC	FastEthernet0/0	192.168.4.2	动态路由； VLAN 路由
		FastEthernet0/1.1	192.168.14.1	
		FastEthernet0/1.2	192.168.15.1	
		FastEthernet0/1.3	192.168.16.1	
		FastEthernet0/1.4	192.168.17.1	
	LiveAreaSwitchA			VTP VLAN 管理； STP 负载分担
	LiveAreaSwitchB			
宿舍区 3	DormArea3Switch			终端接入
宿舍区 4	DormArea4Switch			终端接入
宿舍区 5	DormArea5Switch			终端接入
宿舍区 6	DormArea6Switch			终端接入
	MainAreaRouterFR			帧中继接入
	BranchAreaRouterFR			帧中继接入
	LiveAreaRouterFR			帧中继接入
	FrameRealySwitch			帧中继交换

11.3 需 求 配 置

11.3.1 帧中继网络

1. MainAreaRouterFR

```
interface Serial1/0
   ip address 192.168.1.1 255.255.255.0
   encapsulation frame - relay
```

2. BranchAreaRouterFR

```
interface Serial1/0
   ip address 192.168.1.2 255.255.255.0
   encapsulation frame - relay
```

3. LiveAreaRouterFR

```
interface Serial1/0
```

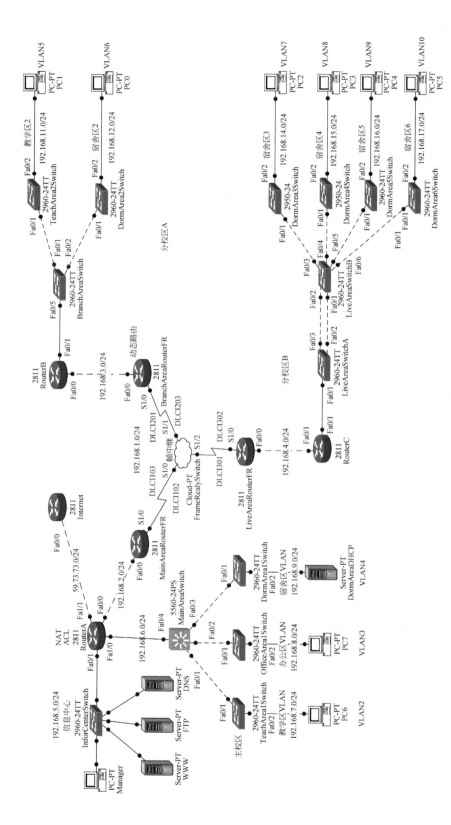

图 11.1 拓扑图

```
ip address 192.168.1.3 255.255.255.0
encapsulation frame-relay
```

4. FrameRealySwitch

```
frame-relay switching

interface Serial1/0
  no ip address
  encapsulation frame-relay
  frame-relay intf-type dce
  frame-relay route 102 interface Serial1/1 201
  frame-relay route 103 interface Serial1/2 301

interface Serial1/1
  no ip address
  encapsulation frame-relay
  frame-relay intf-type dce
  frame-relay route 201 interface Serial1/0 102
  frame-relay route 203 interface Serial1/2 302

interface Serial1/2
  no ip address
  encapsulation frame-relay
  frame-relay intf-type dce
  frame-relay route 301 interface Serial1/0 103
  frame-relay route 302 interface Serial1/1 203
```

11.3.2 访问控制

```
RouterA:
access-list 100 permit tcp 192.168.0.0 0.0.255.255 host 192.168.5.3 eq ftp
access-list 100 permit udp 192.168.0.0 0.0.255.255 host 192.168.5.4 eq domain
access-list 100 permit tcp any host 192.168.5.2 eq www

interface fastethernet0/1
ip access-group 100 out
```

11.3.3 网络地址转换

```
RouterA:
ip nat pool nat 59.73.73.2 59.73.73.254 netmask 255.255.255.0
ip nat inside source list 1 pool nat
access-list 1 permit 192.168.0.0 0.0.255.255

interface fastethernet 0/0
ip nat inside

interface fastethernet 1/0
in nat inside
```

```
interface fastethernet 1/1
ip nat outside
```

11.3.4 动态地址分配

1. RouterB

```
ip dhcp excluded-address 192.168.11.1
ip dhcp excluded-address 192.168.12.1

ip dhcp pool p1
network 192.168.11.0 255.255.255.0
default-router 192.168.11.1
dns-server 192.168.5.4

ip dhcp pool p2
  network 192.168.12.0 255.255.255.0
  default-router 192.168.12.1
  dns-server 192.168.5.4
```

2. RouterC

```
ip dhcp excluded-address 192.168.14.1
ip dhcp excluded-address 192.168.15.1
ip dhcp excluded-address 192.168.16.1
ip dhcp excluded-address 192.168.17.1

ip dhcp pool p3
  network 192.168.14.0 255.255.255.0
  default-router 192.168.14.1
  dns-server 192.168.5.4

ip dhcp pool p4
  network 192.168.15.0 255.255.255.0
  default-router 192.168.15.1
  dns-server 192.168.5.4

ip dhcp pool p5
  network 192.168.16.0 255.255.255.0
  default-router 192.168.16.1
  dns-server 192.168.5.4

ip dhcp pool p6
  network 192.168.17.0 255.255.255.0
  default-router 192.168.17.1
  dns-server 192.168.5.4
```

11.3.5 动静态路由

1. MainAreaRouterFR

```
router rip
```

```
    network 192.168.1.0
    network 192.168.2.0
```

2. BranchAreaRouterFR

```
router rip
    network 192.168.1.0
    network 192.168.3.0
```

3. LiveAreaRouterFR

```
router rip
    network 192.168.1.0
    network 192.168.4.0
```

4. RouterA

```
ip route 0.0.0.0 0.0.0.0 59.73.73.2

router rip
    network 192.168.2.0
    network 192.168.5.0
    network 192.168.6.0
```

5. RouterB

```
router rip
    network 192.168.3.0
    network 192.168.11.0
    network 192.168.12.0
```

6. RouterC

```
router rip
    network 192.168.4.0
    network 192.168.14.0
    network 192.168.15.0
    network 192.168.16.0
    network 192.168.17.0
```

7. MainAreaSwitch

```
router rip
    network 192.168.6.0
    network 192.168.7.0
    network 192.168.8.0
    network 192.168.9.0
```

11.3.6 VLAN 间路由

1. MainAreaSwitch

```
vlan 2
name TeachArea1Vlan
```

```
vlan 3
name OfficeArea1Vlan

vlan 4
name DormArea1Vlan

ip routing

interface FastEthernet0/1
    switchport access vlan 2
    switchport mode access

interface FastEthernet0/2
    switchport access vlan 3
    switchport mode access

interface FastEthernet0/3
    switchport access vlan 4
    switchport mode access

interface Vlan2
    ip address 192.168.7.1 255.255.255.0

interface Vlan3
    ip address 192.168.8.1 255.255.255.0

interface Vlan4
    ip address 192.168.9.1 255.255.255.0
```

2. TeachArea1Switch

```
vlan 2
name TeachArea1Vlan

interface range FastEthernet0/1 - 24
    switchport access vlan 2
    switchport mode access
```

3. OfficeArea1Switch

```
vlan 3
name OfficeArea1Switch

interface range FastEthernet0/1 - 24
    switchport access vlan 3
    switchport mode access
```

4. DormArea1Switch

```
vlan 4
name DormArea1Switch
```

```
interface range FastEthernet0/1 - 24
  switchport access vlan 4
  switchport mode access
```

5. RouterB

```
vlan 5
name TeachArea2Switch

vlan 6
name DormArea2Switch

interface FastEthernet0/1.1
  encapsulation dot1Q 5
  ip address 192.168.11.1 255.255.255.0

interface FastEthernet0/1.2
  encapsulation dot1Q 6
  ip address 192.168.12.1 255.255.255.0
```

6. BranchAreaSwitch

```
interface FastEthernet0/1
  switchport access vlan 5
  switchport mode access

interface FastEthernet0/2
  switchport access vlan 6
  switchport mode access

interface FastEthernet0/5
  switchport mode trunk
```

7. TeachArea2Switch

```
interface range FastEthernet0/1 - 24
  switchport access vlan 5
  switchport mode access
```

8. DormArea2Switch

```
interface range FastEthernet0/1 - 24
  switchport access vlan 5
  switchport mode access
```

11.3.7 STP 负载分担

LiveAreaSwitchB：

```
interface FastEthernet0/1
spanning - tree vlan 7 - 8 port - priority 16
interface FastEthernet0/2
spanning - tree vlan 9 - 10 port - priority 16
```

11.4 设备配置

1. MainAreaRouterFR

```
interface FastEthernet0/0
  ip address 192.168.2.2 255.255.255.0

interface Serial1/0
  ip address 192.168.1.1 255.255.255.0
  encapsulation frame-relay

router rip
  network 192.168.1.0
  network 192.168.2.0
```

2. BranchAreaRouterFR

```
interface FastEthernet0/0
  ip address 192.168.3.1 255.255.255.0

interface Serial1/0
  ip address 192.168.1.2 255.255.255.0
  encapsulation frame-relay

router rip
  network 192.168.1.0
  network 192.168.3.0
```

3. LiveAreaRouterFR

```
interface FastEthernet0/0
  ip address 192.168.4.1 255.255.255.0

interface Serial1/0
  ip address 192.168.1.3 255.255.255.0
  encapsulation frame-relay

router rip
  network 192.168.1.0
  network 192.168.4.0
```

4. FrameRealySwitch

```
frame-relay switching

interface Serial1/0
  no ip address
  encapsulation frame-relay
  frame-relay intf-type dce
  frame-relay route 102 interface Serial1/1 201
  frame-relay route 103 interface Serial1/2 301
```

```
interface Serial1/1
  no ip address
  encapsulation frame-relay
  frame-relay intf-type dce
  frame-relay route 201 interface Serial1/0 102
  frame-relay route 203 interface Serial1/2 302

interface Serial1/2
  no ip address
  encapsulation frame-relay
  frame-relay intf-type dce
  frame-relay route 301 interface Serial1/0 103
  frame-relay route 302 interface Serial1/1 203
```

5. RouterA

```
interface FastEthernet0/0
   ip address 192.168.2.1 255.255.255.0

interface FastEthernet0/1
   ip address 192.168.5.1 255.255.255.0
   ip access-group 100 out

interface FastEthernet1/0
   ip address 192.168.6.1 255.255.255.0

interface FastEthernet1/1
   ip address 59.73.73.1 255.255.255.0

router rip
   network 59.0.0.0
   network 192.168.2.0
   network 192.168.5.0
   network 192.168.6.0

access-list 100 permit tcp 192.168.0.0 0.0.255.255 host 192.168.5.3 eq ftp
access-list 100 permit udp 192.168.0.0 0.0.255.255 host 192.168.5.4 eq domain
access-list 100 permit tcp any host 192.168.5.2 eq www

access-list 1 permit 192.168.0.0 0.0.255.255
ip nat pool nat 59.73.73.3 59.73.73.254 netmask 255.255.255.0
ip nat inside source list 1 pool nat

ip route 0.0.0.0 0.0.0.0 59.73.73.2
```

6. RouterB

```
ip dhcp excluded-address 192.168.11.1
ip dhcp excluded-address 192.168.12.1

ip dhcp pool p1
```

```
  network 192.168.11.0 255.255.255.0
  default-router 192.168.11.1
  dns-server 192.168.5.4

ip dhcp pool p2
  network 192.168.12.0 255.255.255.0
  default-router 192.168.12.1
  dns-server 192.168.5.4

interface FastEthernet0/0
  ip address 192.168.3.2 255.255.255.0

interface FastEthernet0/1.1
  encapsulation dot1Q 5
  ip address 192.168.11.1 255.255.255.0

interface FastEthernet0/1.2
  encapsulation dot1Q 6
  ip address 192.168.12.1 255.255.255.0

router rip
  network 192.168.3.0
  network 192.168.11.0
  network 192.168.12.0
```

7. RouterC

```
ip dhcp excluded-address 192.168.14.1
ip dhcp excluded-address 192.168.15.1
ip dhcp excluded-address 192.168.16.1
ip dhcp excluded-address 192.168.17.1

ip dhcp pool p3
  network 192.168.14.0 255.255.255.0
  default-router 192.168.14.1
  dns-server 192.168.5.4

ip dhcp pool p4
  network 192.168.15.0 255.255.255.0
  default-router 192.168.15.1
  dns-server 192.168.5.4

ip dhcp pool p5
  network 192.168.16.0 255.255.255.0
  default-router 192.168.16.1
  dns-server 192.168.5.4

ip dhcp pool p6
  network 192.168.17.0 255.255.255.0
  default-router 192.168.17.1
  dns-server 192.168.5.4
```

```
interface FastEthernet0/0
   ip address 192.168.4.2 255.255.255.0

interface FastEthernet0/1.1
   encapsulation dot1Q 7
   ip address 192.168.14.1 255.255.255.0

interface FastEthernet0/1.2
   encapsulation dot1Q 8
   ip address 192.168.15.1 255.255.255.0

interface FastEthernet0/1.3
   encapsulation dot1Q 9
   ip address 192.168.16.1 255.255.255.0

interface FastEthernet0/1.4
   encapsulation dot1Q 10
   ip address 192.168.17.1 255.255.255.0

router rip
   network 192.168.4.0
   network 192.168.14.0
   network 192.168.15.0
   network 192.168.16.0
   network 192.168.17.0
```

8. MainAreaSwitch

```
ip routing

interface FastEthernet0/1
   switchport access vlan 2
   switchport mode access

interface FastEthernet0/2
   switchport access vlan 3
   switchport mode access

interface FastEthernet0/3
   switchport access vlan 4
   switchport mode access

vlan 2
name TeachArea1Vlan

vlan 3
name OfficeArea1Vlan

vlan 4
name DormArea1Vlan
```

```
interface Vlan1
   ip address 192.168.6.2 255.255.255.0

interface Vlan2
   ip address 192.168.7.1 255.255.255.0

interface Vlan3
   ip address 192.168.8.1 255.255.255.0

interface Vlan4
   ip address 192.168.9.1 255.255.255.0

router rip
   network 192.168.6.0
   network 192.168.7.0
   network 192.168.8.0
   network 192.168.9.0
```

9. TeachArea1Switch

```
vlan 2
name TeachArea1Vlan

interface range FastEthernet0/1 – 24
   switchport access vlan 2
   switchport mode access
```

10. OfficeArea1Switch

```
vlan 3
name OfficeArea1Vlan

interface range FastEthernet0/1 – 24
   switchport access vlan 3
   switchport mode access
```

11. DormArea1Switch

```
vlan 4
name DormArea1Vlan

interface range FastEthernet0/1 – 24
   switchport access vlan 4
   switchport mode access
```

12. BranchAreaSwitch

```
vlan 5
name TeachArea2

vlan 6
name DormArea2
```

```
interface FastEthernet0/1
  switchport access vlan 5
  switchport mode access

interface FastEthernet0/2
  switchport access vlan 6
  switchport mode access

interface FastEthernet0/5
  switchport mode trunk
```

13. TeachArea2Switch

```
vlan 5
name TeachArea2

interface range FastEthernet0/1 - 24
  switchport access vlan 5
  switchport mode access
```

14. DormArea2Switch

```
vlan 6
name DormArea2

interface range FastEthernet0/1 - 24
  switchport access vlan 6
  switchport mode access
```

15. LiveAreaSwitchA

```
interface FastEthernet0/1
  switchport mode trunk

interface FastEthernet0/2
  switchport mode trunk

interface FastEthernet0/3
  switchport mode trunk

vtp domain LiveArea
vtp mode server
vtp password cisco
vtp version 2

vlan 7
name DormArea3

vlan 8
name DormArea4

vlan 9
```

name DormArea5

vlan 10
name DormArea6

16. LiveAreaSwitchB

interface FastEthernet0/1
  switchport mode trunk
  spanning-tree vlan 7-8 port-priority 16

interface FastEthernet0/2
  switchport mode trunk
  spanning-tree vlan 9-10 port-priority 16

interface FastEthernet0/3
  switchport access vlan 7
  switchport mode access

interface FastEthernet0/4
  switchport access vlan 8
  switchport mode access

interface FastEthernet0/5
  switchport access vlan 9
  switchport mode access

interface FastEthernet0/6
  switchport access vlan 10
  switchport mode access

vtp domain LiveArea
vtp mode client
vtp password cisco
vtp version 2

17. DormArea3Switch

interface range FastEthernet0/1-24
  switchport access vlan 7
  switchport mode access

18. DormArea4Switch

interface range FastEthernet0/1-24
  switchport access vlan 8
  switchport mode access

19. DormArea5Switch

interface range FastEthernet0/1-24
  switchport access vlan 9
  switchport mode access

20. DormArea6Switch

```
interface range FastEthernet0/1 - 24
  switchport access vlan 10
  switchport mode access
```

21. Internet

```
interface FastEthernet0/0
  ip address 59.73.73.2 255.255.255.0
```

参 考 文 献

[1] 吴建胜.路由交换技术.北京:清华大学出版社,2010.
[2] 格拉齐亚尼(Graziani R).思科网络技术学院教程 CCNA Exploration:路由协议和概念.北京:人民邮电出版社,2009.
[3] Bob Vachon,Rick Graziani.思科网络技术学院教程 CCNA Exploration:接入 WAN.北京:人民邮电出版社,2009.
[4] Wayne Lewis.思科网络技术学院教程 CCNA Exploration:LAN 交换无线.北京:人民邮电出版社,2009.
[5] Allan Reid.思科网络技术学院教程 CCNA 4 广域网技术.北京:人民邮电出版社,2008.
[6] 刘易斯(Wayne Lewis).思科网络技术学院教程 CCNA 3 交换基础与中级路由.北京:人民邮电出版社,2008.
[7] Todd Lammle.CCNA 学习指南(640-802)(第 7 版).北京:人民邮电出版社,2012.
[8] 梁广民.思科网络实验室路由、交换实验指南(第 2 版).北京:电子工业出版社,2013.
[9] 王隆杰.思科网络实验室 CCNP(交换技术)实验指南.北京:电子工业出版社,2012.
[10] 梁广民,王隆杰.思科网络实验室 CCNP(路由技术)实验指南.北京:电子工业出版社,2012.
[11] 孙兴华,张晓.网络工程实践教程:基于 Cisco 路由器与交换机.北京:北京大学出版社,2010.

图书资源支持

感谢您一直以来对清华版图书的支持和爱护。为了配合本书的使用,本书提供配套的资源,有需求的读者请扫描下方的"书圈"微信公众号二维码,在图书专区下载,也可以拨打电话或发送电子邮件咨询。

如果您在使用本书的过程中遇到了什么问题,或者有相关图书出版计划,也请您发邮件告诉我们,以便我们更好地为您服务。

我们的联系方式:

地　　址:北京市海淀区双清路学研大厦A座714

邮　　编:100084

电　　话:010-83470236　010-83470237

客服邮箱:2301891038@qq.com

QQ:2301891038(请写明您的单位和姓名)

资源下载:关注公众号"书圈"下载配套资源。

书　圈

获取最新书目

观看课程直播